Nonconvex radially symmetric variational problems

Dissertation
von Stefan Krömer

eingereicht beim
Institut für Mathematik
der Mathematisch–Naturwissenschaftlichen Fakultät
der Universität Augsburg

im September 2005

Augsburger Schriften zur Mathematik, Physik und Informatik
Band 11

herausgegeben von:
Professor Dr. F. Pukelsheim
Professor Dr. W. Reif
Professor Dr. D. Vollhardt

Bibliografische Information Der Deutschen Bibliothek

Die Deutsche Bibliothek verzeichnet diese Publikation in der Deutschen Nationalbibliografie; detaillierte bibliografische Daten sind im Internet über http://dnb.ddb.de abrufbar.

©Copyright Logos Verlag Berlin 2005
Alle Rechte vorbehalten.

ISBN 3-8325-1240-3
ISSN 1611-4256

Logos Verlag Berlin
Comeniushof, Gubener Str. 47,
10243 Berlin
Tel.: +49 030 42 85 10 90
Fax: +49 030 42 85 10 92
INTERNET: http://www.logos-verlag.de

Vorwort

Nichtkonvexe Variationsprobleme stellen aus vielen Gründen eine mathematische Herausforderung dar. Selbst wenn das Funktional nach unten beschränkt und koerziv ist, ist nicht klar, ob ein Minimierer existiert. Es war Weierstraß, der seine Zeitgenossen auf diesen subtilen Umstand hinwies und durch ein Gegenbeispiel das Dirichletsche Prinzip ins Wanken brachte. (Später wurde dies von Hilbert rehabilitiert.) Selbst die Existenz kritischer Punkte ist unklar, da die Euler–Lagrange Gleichung bei fehlender Konvexität nicht elliptisch ist, sondern offensichtlich den Typ wechselt. Randwertprobleme vom wechselnden Typ entziehen sich aber jeder allgemeinen Theorie.

Trotz aller Schwierigkeiten müssen sich die Mathematiker aber den nichtkonvexen Variationsproblemen stellen, da viele physikalische Modelle z.B. für Phasenübergänge in Materialien durch ein nichtkonvexes Energiefunktional dargestellt werden. Herr Krömer beschreibt diese in seiner Einleitung, geht später aber nicht mehr auf die physikalische Anwendung ein. Sie rechtfertigen und motivieren in ausreichender Weise seine Mathematik. Für eindimensionale Modelle, d.h. für Energieintegrale über einem Intervall, gibt es bereits viele spezielle und auch allgemeine Resultate zur Existenz bzw. Nichtexistenz von Minimierern. Für höherdimensionale Integrale indessen ist wenig bekannt; es überwiegen die Beispiele für ein zunehmend oszillatorisches Verhalten von Minimalfolgen, welches z.B. eine kristalline Mikrostruktur beschreibt, die die minimale Energie nicht annehmen kann. Einen Kompromiss zwischen erfolgreichen eindimensionalen Techniken und einer Höherdimensionalität des Problems stellt die Radialsymmetrie dar: Das Integral radialsymmetrischer Funktionen über einer Kugel ist ein eindimensionales Integral über den Radius. Die Singularität im Nullpunkt lässt aber eine Übertragung eindimensionaler Resultate nicht unmittelbar zu. Dazu bedarf es zusätzlicher, z.T. subtiler Überlegungen.
Ein bewährtes Hilfsmittel der Variationsrechnung ist eine konvexe Relaxation, z.B. durch die Konvexifizierung des Potentials oder durch die Addition eines konvexen Terms höherer, hier zweiter Ordnung, in der Ableitung. Die Kleinheit dieses Terms wird durch den Koeffizienten ϵ gewährleistet, der dann in einem singulären Grenzübergang gegen Null strebt und als Grenzfunktional das ursprüngliche wieder herstellt. Die Dissertation umfasst zwei Teile: Das Studium des nichtkonvexen Grenzfunktionals und das des gestörten konvexen Funktionals mit singulärem Grenzübergang. Das Grundgebiet ist eine Kugel.

Das erste interessante Resultat ist die Existenz radialsymmetrischer Minimierer im konvexen Fall. Dieses wird verschärft durch die Aussage, dass unter vernünftigen Voraussetzungen alle Minimierer über einer Kugel radialsymmetrisch sind. Dies wiederum führt zu einer tiefen Aussage im nichtkonvexen Fall: Die radialsymmetrischen Minimierer des konvexifizierten Funktionals sind auch Minimierer des nichtkonvexen Funktionals, was zu einer Existenzaussage über Minimierer für eine neue Klasse nichtkonvexer Variationsprobleme führt. Dass die Radialsymmetrie dabei eine entscheidende Rolle spielt, zeigt ein Gegenbeispiel. Der erste Teil der Arbeit schließt mit qualitativen Aussagen über radialsymmetrische kritische Punkte, d.h. über Lösungen der Euler–Lagrange Gleichung. Dazu gehört eine Eindeutigkeitsaussage über gewisse kritische Punkte, zu denen auch die Minimierer gehören. Die Techniken sind subtil und verwenden u.a. eine verfeinerte Version des Hopfschen Maximumprinzips.

Im zweiten Teil wird das singulär gestörte Funktional mit Hilfe einer globalen Verzweigungsanalyse untersucht. Dazu wird das Funktional so homotopiert, dass für gewisse Werte des Homotopieparameters radialsymmetrische Lösungen der Euler–Lagrange Gleichung von der trivialen Lösunge verzweigen. Die topologische Methode von Rabinowitz garantiert die Existenz globaler Lösungszweige, und die nichttriviale Aufgabe besteht nun darin, durch eine globale qualitative Analyse den Verlauf der Zweige zu bestimmen. Der erste Schritt besteht in einer a priori Abschätzung, welche die globalen Zweige zwingt, wieder zur trivialen Lösung zurückzukehren. Der zweite entscheidende ist die Bestimmung des Zielpunktes. Existenzaussagen helfen da nicht weiter – nur qualitative Eigenschaften der Lösungen, welche einen unter vielen möglichen Verzweigungspunkten auf der trivialen Achse auswählen. Diese Erkennungsmerkmale sind die Vorzeichen sowohl der Lösung als auch einiger ausgewählter Ableitungen. Dass diese auf Lösungskontinua (d.h. unter Homotopien) erhalten bleiben, ist eine Konsequenz von Maximumprinzipien, die allerdings für die Euler–Lagrange Gleichung vierter Ordnung (das singulär gestörte Funktional ist zweiter Ordnung) nicht offensichtlich sind. Die Quintessenz ist, dass es einen nichttrivialen radialsymmetrischen Lösungszweig der Euler–Lagrange Gleichung gibt, welcher zwei Verzweigungspunkte verbindet. Es sind Lösungen der Euler–Lagrange Gleichungen des homotopierten Funktionals, und die ganze Analyse ist deshalb sinnvoll, weil das ursprüngliche (singulär gestörte) Funktional zum Bereich des Homotopieparameters zwischen den Verzweigungspunkten gehört. Herr Krömer hat damit bewiesen, dass das singulär gestörte Funktional einen kritischen Punkt mit ausgewählten Eigenschaften besitzt. Diese qualitativen Eigenschaften sind es auch, die den singulären Grenzübergang zum nichtkonvexen Funktional und seiner Euler–Lagrange Gleichung zweiter Ordnung erlauben. Diese Kompaktheitsaussagen sind äußerst subtil und hängen von den Vorzeichen der radialsymmetrischen Funktionen und der ausgewählten Ableitungen ab. Der singuläre Grenzwert ist dann ein kritischer Punkt des nichtkonvexen Funktionals erster Ordnung.

Damit ist die Arbeit noch nicht ganz getan: Dieser kritische Punkt ist der eindeutig bestimmte Minimierer! Es ist weniger die Existenz des Minimierers (die hat Herr Krömer schon im ersten Teil der Arbeit bewiesen) als die Eigenschaft, singulärer Grenzwert zu sein, welche dem zweiten Teil seine Bedeutung verleiht: Numerische Verfahren greifen nur für die elliptische Euler–Lagrange Gleichung des singulär gestörten Funktionals. Eine Pfadverfolgung mittels des eingeführten Homotopieparameters, ausgehend von der trivialen Lösung, liefert eine Lösung, welche für kleine ϵ den Minimierer gut approximiert. Dies liefert ein praktikables Verfahren der numerischen Simulation des Minimierers eines nichtkonvexen Variationsproblems.

Augsburg, den 7. April 2006

Prof. Dr. Hansjörg Kielhöfer
Universität Augsburg

Abstract

The energy functional

$$E_0(u) := \int_\Omega W(\nabla u) + G(u)\, dx$$

is discussed, where u is a real valued function over $\Omega \subset \mathbb{R}^N$ which vanishes on the boundary and W is nonconvex. The functional is assumed to be radially symmetric in the sense that Ω is a ball and W only depends on $|\nabla u|$. In the first chapter, existence and radial symmetry of a global minimizer are shown with an approach based on convex relaxation. This is done without assuming that G is convex, thus extending a result of Cellina and Perrotta [7]. Moreover, an example without radial symmetry is given where a global minimizer does not exist despite the fact that G is strictly concave. In the second chapter, the functional is regularized by adding the term $\epsilon(\Delta u)^2$ in the integrand, where $\epsilon > 0$ is a small parameter. As in the corresponding one–dimensional case ($N = 1$, cf. [23]), the corresponding Euler–Lagrange equation is analyzed as a bifurcation problem. The existence of a global branch of positive radially symmetric solutions is shown, using the degree of Leray and Schauder, the maximum principle and an a–priori estimate in $W^{1,\infty}$ which is uniform in ϵ as the main tools. Those solutions converge to nontrivial solutions of the Euler-Lagrange equations of E_0 as ϵ goes to zero, and the limits coincide with the global minimizer of E_0 for suitable W and G.

MSC: 49J10, 49J45, 35B25, 35B32, 35J40

Contents

Introduction

This thesis is devoted to the study of the variational problem arising from the energy functional

$$E_\epsilon(u) := \int_\Omega \left[\frac{\epsilon}{2}(\Delta u)^2 + W(\nabla u) + G(u)\right] dx, \tag{1}$$

both for $\epsilon = 0$ and $\epsilon > 0$, where u is a scalar field on $\Omega \subset \mathbb{R}^N$ $(N \geq 2)$ which vanishes on the boundary. We will restrict ourselves mostly to the radially symmetric case, i. e. when Ω is a ball and W is invariant under rotations in the sense that W only depends on the euclidean norm of its argument. Typical examples for the functions W and G considered are $W(\xi) := (|\xi|^2 - 1)^2$ and $G(\mu) := -\mu^2$. The primary qualitative features of W are that it is nonconvex, coercive and frame indifferent in the sense that it only depends on the euclidean norm of its argument. As to G, we only exploit the fact that it is strictly decreasing on $\mathbb{R}^+$.

The energy E_ϵ can be considered as a toy problem bearing characteristic features closely related to a variety of physical models. For an overview, we refer to [21]. The first term of E_ϵ is usually interpreted as a surface energy, penalizing fast changes of the gradient. However, note that surface energy usually is modeled by a term of the form $|D^2u|^2$ instead of $(\Delta u)^2$. Both terms yield the same Euler-Lagrange equations, but the corresponding natural boundary conditions differ. A close relative of E_ϵ is

$$\int_\Omega (|\nabla u| - 1)^2 + \epsilon \left|\operatorname{div}\left(\frac{\nabla u}{|\nabla u|}\right)\right|^2$$

which models the energy of a so-called smectic-A liquid crystal [26]. In this setting, one typically has $\Omega \subset \mathbb{R}^3$, the domain of the liquid crystal, and $u : \Omega \to \mathbb{R}$ is a suitable transformation of the local mass density of the material. Similar models also emerge in the theory of nonlinear elasticity for thin films attached to a planar substrate $\Omega \subset \mathbb{R}^2$ [29]. If we assume that the in-plane displacements of the film always vanish ("anti-plane shear", which of course is a rather crude simplification), the corresponding elastic energy can be modeled as

$$\int_\Omega \epsilon^3 \left((1-\nu)\left|D^2u\right|^2 + \nu\left|\Delta u\right|^2\right) + \epsilon\left(|\nabla u|^2 - \delta\right)^2 dx$$

(qualitatively, omitting some constants). Here, $u : \Omega \to \mathbb{R}$ is the out-of-plane displacement of the film (buckling away from the substrate), $\epsilon > 0$ is the thickness of the (thin) film, $\nu \in (-1, 1/2)$ is Poisson's ratio of the material and the constant $\delta > 0$ is the eigenstrain of the film which takes into account that the film attached to the substrate is not in its relaxed state but actually slightly compressed (and thus trying to buckle out of the plane). For both models one can add terms depending on u, which in the case of the elastic thin film correspond to external forces (such as gravity, for example). Here, note that the prototype of G corresponds to a force pulling the film away from the substrate, a situation in which one intuitively would not expect the development of the microstructures studied in [29]. It is not so surprising that this is related to the compactness of critical points as $\epsilon \to 0$, a question which we address in the second chapter.

Apart from any questions related to similar physical models, the energy E_ϵ also leads to a number of interesting mathematical challenges. In the case $\epsilon = 0$, the main difficulty arises from the fact that W in nonconvex. This prevents an application of the direct methods of the calculus of variations to obtain minimizers, since by lack of convexity the energy lacks weak lower semicontinuity. Also any coercive but nonconvex W (which is smooth enough) has areas where its Hessian $D\nabla W$ is indefinite and thus the corresponding Euler-Lagrange equation switches its type from elliptic to hyperbolic, depending on the value of ∇u.

Abundant literature addressing the existence and further properties of global minimizers for $N > 1$ is available even without assuming radial symmetry, see [30, 5, 4] for the most general results and [4] for further references. Generalizations for vector-valued u can be found in [31, 32]. However there are still open questions. For example, to ensure existence of a minimizer, the above mentioned papers in particular have to assume that the convex (respectively, quasiconvex) hull W^c of W has the following property:

$$W^c \text{ is affine on any component of the detachment set } \{W^c < W\}. \tag{2}$$

Our radially symmetric prototype example above of course satisfies (2), but no multi-well potential W whose set of global minima consists of a finite number of points verifies it, and even if W is radially symmetric, any nonconvex parts outside the (outermost) sphere of minima are ruled out. The radially symmetric case, where, in particular, Ω is a ball, is studied in [7, 8, 11, 18]. There, (2) can be dropped provided that G is convex and decreasing, a result first stated in [7] with a wrong proof. A corrected proof can be found in [8]. The convexity of G is also exploited to prove the existence of a radially symmetric minimizer. In the first chapter we will show in particular that convexity of G is actually a technical assumption in the sense that it can be dropped if G is of class C^2.

Another natural object of investigation is the "singular" limit as ϵ tends to zero. One interesting open problem in this context is to obtain the Γ-limit of $\epsilon^{-1/2}E_\epsilon$ as $\epsilon \to 0$ when $G \equiv 0$ for general domains Ω. A likely candidate satisfying the Γ-liminf inequality is available on two-dimensional domains for a limited class of functions W of polynomial type, however a proof of the Γ-limsup property is still missing, see [21, 15] and the references therein. Even less is known in higher dimensions [13]. It is interesting to note that the difficulty of this problem effectively arises from the radial symmetry of W, or more generally speaking, from the fact that the set of minima of W is a sphere which, in particular, is not a discrete set. We will not go in this direction, however. Actually our attention is focused on a complementary setting, since our assumptions on G ensure the existence of a global minimizer for $\epsilon = 0$, which is even unique in the radially symmetric case (at least for a broad class of W and G, including the prototype). Thus the development of microstructure at the energy level of the global minimizer is not to be expected.

This thesis encompasses two chapters. The first one is concerned with the limit problem corresponding to $\epsilon = 0$. The question of existence of radially symmetric minimizers for nonconvex W is investigated in detail. As a byproduct, the discussion shows that actually all global minimizers are necessarily radially symmetric for a broad class of potentials bearing some of the qualitative properties of our prototype. We also give a non-existence result where W is a (non-radially symmetric) double well, to illustrate the fact that the condition (2) is not purely technical in a higher-dimensional setting lacking radial symmetry. We then continue to derive properties for critical points in general, including a priori bounds and regularity results. The latter are based on the Weierstrass-Erdmann corner conditions, which are of special interest in particular at the center of the ball where radially symmetric solutions behave differently as in the the one-dimensional case. We conclude the chapter with an uniqueness result for critical points within an appropriate class of functions which can be used to identify singular limits as the global minimizers of the limit problem. The second chapter contains results for $\epsilon > 0$ as well as the study of the singular limit process. Our setting entails that $u \equiv 0$ always is a "trivial" critical point, but our interest is essentially focussed on nontrivial critical points. Thus, for fixed ϵ, we study the Euler-Lagrange equations associated to E_ϵ as a global bifurcation problem, with the Leray–Schauder degree and the Hopf's maximum principle as the main tools. Our work is motivated by [23], where the one-dimensional case is studied. The approach based on bifurcation analysis has the benefit that one can obtain a lot of information on the qualitative properties of solutions on branches, which can be exploited to show compactness as ϵ goes to zero. The drawback is that despite the fact we know a lot about some nontrivial solutions, we are not able to show their stability. On the other hand, a purely variational approach would have to face the problem that obtaining qualitative properties of global or local minimizers directly is quite hard, again caused by the fact that the problem is of fourth order (respectively, the energy contains derivatives of second order). That also leads to serious difficulties if one tries to prove compactness in the singular limit because one has less knowledge about the shape of the minimizer – results in that direction are only available if $G \equiv 0$ [15], an assumption which greatly simplifies the limit functional. As to carrying our program, the main problems also arise from the fact that the equation is of fourth order. In particular, this is a serious obstacle for the application of a maximum principle, which we overcome by exploiting the radial symmetry. Of course this (technically) prevents a straightforward generalization to non-symmetric problems. Our second interesting result based on radial symmetry is an a-priori estimate in $W^{1,\infty}$ for solutions of the Euler-Lagrange equation, which is uniform in ϵ. It proves very useful both for bounding global continua of solutions for fixed ϵ and for studying the limit as $\epsilon \to 0$. The latter is exposed in the last part of the second chapter. There, we show compactness in $W^{1,P}$ ($P \in (1, \infty)$ arbitrary) of the solutions obtained by the global bifurcation technique as $\epsilon \to 0$. A result which ensures that the limit is nontrivial concludes our investigation.

Acknowledgements

I am indebted to my supervisor Prof. Hansjörg Kielhöfer for giving me the opportunity to write this thesis. I appreciate his constant interest in my work as well as the numerous stimulating discussions and suggestions on the subject. Furthermore, I want to express my gratitude to Markus Lilli and Prof. Timothy Healey for their feedback, and to Ulrich Miller, who provided numerical tools and results for related problems which gave valuable hints on some of the phenomena to be expected in situations without radial symmetry. The financial support I received from the Graduiertenkolleg "Nichtlineare Probleme in Analysis, Geometrie und Physik" at the University of Augsburg is also gratefully acknowledged.

Basic notation

The letters C and c each are used for multiple real constants which do not necessarily have the same value. Given two vectors $\xi, \eta \in \mathbb{R}^N$, $\xi \cdot \eta$ or $\xi\eta$ denotes their euclidean scalar product. The euclidean norm in $\mathbb{R}^N$ as well as the absolute value in $\mathbb{R}$ are denoted by $|\cdot|$. The symbol $\|\cdot\|$ is used for norms in function spaces, where the corresponding space will be given in the index, for example $\|\cdot\|_{L^p(\Omega)}$. Sobolev spaces of real-valued functions in $L^p(\Omega)$ which are k times weakly differentiable in $L^p(\Omega)$ are denoted by $W^{k,p}(\Omega)$, and $W_0^{k,p}(\Omega) \subset W^{k,p}(\Omega)$ stands for the closure of the set of infinitely times differentiable functions with compact support in Ω (i.e., $C_0^\infty(\Omega)$) with respect to the $W^{k,p}$-norm. Moreover, we say that a function is of class C^k if all derivatives up to k-th order exist in the classical sense and are continuous, whereas $C^k(\Omega)$ is space of function of class C^k on Ω with uniformly bounded derivatives. By $C^{k,\alpha}(\Omega)$, we denote the Hölder space of k times classically differentiable functions with Hölder-continuous derivatives, where $\alpha \in (0,1)$ is the corresponding Hölder exponent. As far as Dirichlet boundary conditions are involved, we use $C_D^{k,\alpha}(\Omega) := \{u \in C^{k,\alpha}(\Omega) \mid u = 0 \text{ on } \partial\Omega\}$ (more precisely, $u(x) \to 0$ as $\operatorname{dist}(x; \partial\Omega) \to 0$), and $C_D^k(\Omega)$ is defined analogously. In the notation for all those spaces we will omit the domain Ω if it is clear from the context. Open balls in $\mathbb{R}^N$ are denoted by $B_R(a)$, where $R > 0$ is the radius and $a \in \mathbb{R}^N$ is the center of the ball. Furthermore, the $N-1$-dimensional sphere is defined as $S^{N-1} := \partial B_1(0) \subset \mathbb{R}^N$, the boundary of the unit ball in $\mathbb{R}^N$. The restriction of a function space over a ball to the set of radially symmetric functions is indicated by the letter "r" in the index, for example, $W_{0,r}^{1,p}(B_R(0)) := \{u \in W_0^{1,p}(B_R(0)) \mid u \text{ is radially symmetric}\}$. In that context, "radially symmetric" means that u is almost everywhere equal to a radially symmetric function. With a slight abuse of notation, we use the same letter both for a radially symmetric function $u : B_R(0) \subset \mathbb{R}^N \to \mathbb{R}$ and its "profile" $u : [0, R) \to \mathbb{R}$ related by $u(x) = u(|x|)$. The derivative of u in radial direction is denoted by $\partial_r u$, respectively, u'.

Chapter 1

The limit functional

In this chapter, we study the functional

$$E_0(u) := \int_\Omega [W(\nabla u) + G(u)]\, dx. \tag{E_0}$$

We consider E_0 as a functional on $W_0^{1,p}(\Omega)$ where Ω is a domain in $\mathbb{R}^N$, $N \geq 2$. Most of our attention is focused on the radially symmetric case, i.e. if Ω is a ball and W is radially symmetric in the sense that it only depends on the norm of ∇u. Here, we restrict ourselves to results obtained by studying E_0 directly, as opposed to the approach via singular perturbation exposed in the second chapter. Both minima and (more general) critical points are discussed, asking the questions of existence, uniqueness and qualitative properties.

1.1 Preliminaries

The functions W and G usually are supposed to satisfy the conditions stated below. We do not give a complete list however; the properties collected here are just the most common ones.

Assumptions on W:

$$\text{(Regularity) } W : \mathbb{R}^N \to \mathbb{R} \text{ is of class } C^0, \tag{W_0}$$

$$\text{(Coercivity) } W(\xi) \geq \nu_1 |\xi|^p - C, \tag{W_1}$$

$$\text{(Growth) } \quad |W(\xi)| \leq \nu_2 |\xi|^p + C,, \tag{W_2}$$

for every $\xi \in \mathbb{R}^N$, where $p > 1$, ν_1, ν_2 and C are positive real constants. For the case of radial symmetry, which we will discuss in detail, we furthermore assume that Ω is a ball and that W is invariant under rotations:

$$\text{(Symmetry)} \quad \begin{array}{l} W(\xi) = \tilde{W}(|\xi|), \text{ where} \\ \tilde{W} : \mathbb{R} \to \mathbb{R} \text{ is an even function of class } C^0. \end{array} \qquad (W_{sym})$$

Note that in particular we do not require W to be convex. If W is nonconvex, the so-called Maxwell points M and $-M$, defined below, are of special interest:

$$M := \max\left\{t \geq 0 \mid \tilde{W}(t) = \min_{s\in\mathbb{R}} \tilde{W}(s)\right\}. \tag{1.1}$$

Another important object in the study of nonconvex $\tilde{W}$ is its convex hull

$$\tilde{W}^c(s) := \sup\left\{V(s) \mid V : \mathbb{R} \to \mathbb{R} \text{ is convex and } V \leq W\right\}, \; s \in \mathbb{R}. \tag{1.2}$$

Note that if $\tilde{W}$ is continuous or of class C^1 then the same holds for $\tilde{W}^c$. Furthermore, $\tilde{W}^c$ is convex and affine on any component of the set where it differs from $\tilde{W}$. The Maxwell points have the property that $\tilde{W}(\pm M) = \tilde{W}^c(\pm M)$.

Remark 1.1.1. If W is replaced by a function $\hat{W}$ of the form $\hat{W}(\xi) = W(\xi) + a\cdot\xi$, where $a \in \mathbb{R}^N$ is an arbitrary fixed vector, then the energy E_0 remains unchanged, by virtue of the Gauss Theorem. In particular, all critical points persist. This invariance can be used for example to treat some cases when W is "skew", as opposed to our assumption (W_{sym}).

Assumptions on G:

$$\textit{(Regularity)} \quad G : \mathbb{R} \to \mathbb{R} \text{ is of class } C^0, \qquad (G_0)$$

$$\textit{(Shape)} \quad \begin{array}{l} G \text{ is strictly decreasing on } [0,\infty) \text{ and} \\ G(\mu) \leq G(-\mu) \text{ whenever } \mu > 0, \end{array} \qquad (G_1)$$

$$\textit{(Growth)} \quad |G(\mu)| \leq \nu_3 |\mu|^{\tilde{p}} + C, \; \tilde{p} < p, \qquad (G_2)$$

for every $\mu \in \mathbb{R}$, where $\tilde{p}$, $C \in \mathbb{R}$ and $\nu_3 \geq 0$ are constants. A first consequence of the conditions given above is the following

Lemma 1.1.2. *(Coercivity of E_0) Assume (W_0)-(W_2), (G_0) and (G_2). Then $E_0 : W_0^{1,p}(\Omega) \to \mathbb{R}$ is well defined and coercive in the sense that*

$$E_0(u) \geq \tilde{\nu}\, \|u\|^p_{W^{1,p}} - \tilde{C}, \tag{1.3}$$

for every $u \in W_0^{1,p}(\Omega)$, where $\tilde{\nu} > 0$ and $\tilde{C}$ are real constants independent of u.

Proof. Using the growth conditions, it is not difficult to show that E_0 is well defined. Furthermore, for $u \in W_0^{1,p}(\Omega)$, by virtue of (W_1), (G_2), Hölder's inequality and Poincaré's inequality we have that

$$\begin{aligned} E_0(u) &\geq \int_\Omega \left[\nu_1 |\nabla u|^p - \nu_3 |u|^{\tilde{p}} - 2C\right] dx \\ &\geq \tilde{\nu}_1 \|u\|^p_{W^{1,p}} - \tilde{\nu}_3 \|u\|^{\tilde{p}}_{W^{1,p}} - 2C, \end{aligned}$$

where $\tilde{\nu}_1$ and $\tilde{\nu}_3$ are positive constants depending on ν_1 and ν_3, respectively, as well as on p, $\tilde{p}$ and $\mathrm{Vol}(\Omega)$. Since $\tilde{p} < p$, this immediately implies (1.3). □

Part of this chapter will contain results which do not only refer to minimizers but to critical points in general. We then have to strengthen the assumptions above in such a way that E_0 can be differentiated with respect to $u \in W_0^{1,p}(\Omega)$:

Assumptions on W:

(Regularity)	$W : \mathbb{R}^N \to \mathbb{R}$ is of class C^1,	(W_0^{EL})
(Coercivity)	$DW(\xi)\xi \geq \nu_1 |\xi|^p - C$,	(W_1^{EL})
(Growth)	$|DW(\xi)| \leq \nu_2 |\xi|^{p-1} + C,$,	(W_2^{EL})

for every $\xi \in \mathbb{R}^N$, where $p > 1$, ν_1, ν_2 and C are positive real constants. In the case of radial symmetry we require

(Symmetry)	$W(\xi) = \tilde{W}(|\xi|)$, where $\tilde{W} : \mathbb{R} \to \mathbb{R}$ is an even function of class C^1.	(W_{sym}^{EL})

Assumptions on G:

(*Regularity*)	$G : \mathbb{R} \to \mathbb{R}$ is of class C^1,	(G_0^{EL})
(*Shape*)	G is strictly decreasing on $[0, \infty)$ and $G(\mu) \leq G(-\mu)$ whenever $\mu > 0$,	(G_1^{EL})
(*Growth*)	$|G'(\mu)| \leq \nu_3 |\mu|^{\tilde{p}-1} + C,\ \tilde{p} < p$,	(G_2^{EL})

1.2 Properties of minimizers in the convex case

In the case of convex W, the functional E_0 is weakly lower semicontinuous, and since it is also coercive by Lemma 1.1.2, E_0 has a minimum by the direct methods in the Calculus of Variations (cf. [12]) in $W_0^{1,p}$. Of course even in this case it is not immediately clear that the minimizer has radial symmetry if W itself is radially symmetric in the sense of (W_{sym}). This section provides several auxiliary results which are employed to show existence and symmetry of minimizers for nonconvex W in Section 1.3. For this purpose, we will apply the assertions below to the relaxed functional where W replaced by its convex hull W^c. As a consequence, we actually could assume that $W = W^c$ within this section. However, the arguments used here do not really exploit convexity of W (although convexity is always sufficient) which guarantees the existence of a minimizer. Thus we prefer to use a more general setting, assuming just those properties of W which are really needed for the proofs and ignoring the fact that in some cases the results might be empty in the sense that they provide properties for minimizers which do not exist. Note however that one does not achieve a true generalization of the case of convex W in this way since a global minimizer for nonconvex W also has to minimize the convexified functional where W is replaced with its convex hull W^c, due to the Relaxation Theorem .

As a first step, we discuss the question of radial symmetry of minimizers, assuming suitable symmetry of W. As mentioned above, the existence of a minimizer is quite clear (as long as W is convex), thus we only have to show that one (or, preferably, every) minimizer is radially symmetric. This can be done by constructing radially symmetric functions in a suitable way from a given, possibly non-symmetric minimizer. The following lemma ensures sufficient regularity of those functions:

Lemma 1.2.1. *Let $u \in W^{1,p}(B_R(0))$ with a number $p \in [1, \infty)$. Then, for almost every direction $\theta \in S^{N-1} := \partial B_1(0) \subset \mathbb{R}^N$, the radially symmetric function*

$$u_\theta : B_R(0) \to \mathbb{R}, \ u_\theta(x) := u(|x|\,\theta) \tag{1.4}$$

is an element of $W^{1,p}(B_R(0))$. If $u \in W_0^{1,p}(B_R(0))$, then we also have $u_\theta \in W_0^{1,p}(B_R(0))$ for a. e. $\theta \in S^{N-1}$. In any case,

$$\nabla u_\theta(r\psi) = (\theta \cdot \nabla u(r\theta))\,\psi, \tag{1.5}$$

and in particular,

$$|\nabla u_\theta(r\psi)| \leq |\nabla u(r\theta)|\,, \tag{1.6}$$

for almost every $r \in (0, R)$, $\theta \in S^{N-1}$ and every $\psi \in S^{N-1}$.

Proof. We will only give the proof for the case if u is an element of $W_0^{1,p}(B_R(0))$, the modifications for $u \in W^{1,p}(B_R(0))$ are obvious (note that in this case approximation by smooth functions on $\overline{B_R(0)}$ is possible, too, since the boundary of the ball is regular enough).

Since u is an element of $W_0^{1,p}(B_R(0))$, it can be approximated with a sequence $u^{(k)} \in C_0^\infty(B_R(0))$, $k \in \mathbb{N}$ such that $u^{(k)} \to u$ in $W^{1,p}$. Obviously the functions $u_\theta^{(k)}$ are elements of $C^\infty(B_R(0) \setminus \{0\}) \cap C(B_R(0))$ and vanish in a vicinity of $\partial B_R(0)$, for every $k \in \mathbb{N}$ and every direction $\theta \in S^{N-1}$. Since for fixed k $\nabla u^{(k)}(0)$ is finite, the discontinuity of the gradient of $u_\theta^{(k)}$ at 0 is not too bad, so that we also have $u_\theta^{(k)} \in W_0^{1,p}(B_R(0))$. Therefore, $u_\theta \in W_0^{1,p}(B_R(0))$ follows once we show that $u_\theta^{(k)} \to u_\theta$ in L^p as $k \to \infty$ and that $(\nabla u_\theta^{(k)})_k \to \nabla u_\theta$ in L^p for almost every $\theta \in S^{N-1}$, where ∇u_θ is given by (1.5). This can be observed in the following way: By introducing radial coordinates, using Fubini's Theorem, we have

$$\begin{aligned}
&\int_{S^{N-1}} \int_{B_R(0)} \left| u_\theta^{(k)} - u_\theta \right|^p dx d\theta \\
&= \int_{S^{N-1}} \int_{S^{N-1}} \int_0^R \left| u_\theta^{(k)}(r\psi) - u_\theta(r\psi) \right|^p r^{N-1} dr d\psi d\theta \\
&= \int_{S^{N-1}} \int_{S^{N-1}} \int_0^R \left| u_\theta^{(k)}(r\theta) - u_\theta(r\theta) \right|^p r^{N-1} dr d\psi d\theta \\
&\qquad \text{since } u_\theta^{(k)} \text{ and } u_\theta \text{ are radially symmetric} \\
&= \int_{S^{N-1}} \int_{S^{N-1}} \int_0^R \left| u^{(k)}(r\theta) - u(r\theta) \right|^p r^{N-1} dr d\theta d\psi \\
&= \mathrm{Vol}_{N-1}(S^{N-1}) \int_{B_R(0)} \left| u^{(k)} - u \right|^p dx \\
&\to 0 \text{ as } k \to \infty.
\end{aligned}$$

Thus, $u_\theta^{(k)} \to u_\theta$ in L^p for a. e. θ. By a similar calculation we obtain

$$\begin{aligned}
&\int_{S^{N-1}} \int_{B_R(0)} \left| \nabla u_\theta^{(k)}(r\psi) - (\theta \cdot \nabla u(r\theta))\, \psi \right|^p dx d\theta \\
&\qquad \le \mathrm{Vol}_{N-1}(S^{N-1}) \int_{B_R(0)} \left| \nabla u^{(k)} - \nabla u \right|^p dx,
\end{aligned} \tag{1.7}$$

where $r := |x|$ and $\psi := x/|x|$, using that

$$\begin{aligned}
\left| \nabla u_\theta^{(k)}(r\psi) - (\theta \cdot \nabla u(r\theta))\, \psi \right| &= \left| \left(\theta \cdot \nabla u^{(k)}(r\theta)\right) \psi - (\theta \cdot \nabla u(r\theta))\, \psi \right| \\
&\le \left| \nabla u^{(k)}(r\theta) - \nabla u(r\theta) \right|,
\end{aligned}$$

for every $r \in (0, R)$ and $\psi, \theta \in S^{N-1}$. Since $\nabla u^{(k)}$ converges to ∇u in L^p, (1.7) immediately entails (1.5). □

Remark 1.2.2. Even stronger results about regularity properties of the traces of a Sobolev function on a set of parallel lines which form a partition of the domain can be found in [16]. However the results presented there are not directly applicable in the situation of the lemma above because the lines in radial direction meet at the origin, thus behaving (mildly) singular.

As an technical tool in order to prove the symmetry of a whole group of minimizers (even all for suitable W), we need the following elementary characterization of radially symmetric functions:

Lemma 1.2.3. *Assume that $u \in W^{1,1}_{loc}(B_R(0))$ satisfies*

$$\nabla u(x) = \lambda(x)x \quad \textit{for a. e. } x \in B_R(0), \tag{1.8}$$

where $\lambda = \lambda(x) \in \mathbb{R}$ is a scalar factor. Then u is radially symmetric.

Proof. Using approximation with smooth functions and Fubini's Theorem, it is not difficult to show that the trace functions $\theta \mapsto u_r(\theta) := u(r\theta)$, $S^{N-1} \to \mathbb{R}$, belong to $W^{1,1}_{\text{loc}}(S^{N-1})$ for almost every $r \in (0, R)$. Furthermore,

$$Du_r(\theta)h = rDu(r\theta)h \text{ for } h \in T_\theta S^{N-1}.$$

Due to (1.8),

$$Du_r(\theta)h = r^2\lambda(r\theta)(\theta \cdot h) = 0,$$

since the tangential vector $h \in T_\theta S^{N-1} \subset \mathbb{R}^N$ is always orthogonal to θ. Thus u_r is constant on S^{N-1} for almost every r. Accordingly, u is constant on the spheres $\partial B_r(0)$ for almost every $r \in (0, R)$, which entails radial symmetry. $\square$

With the aid of Lemma 1.2.1 we now can show radial symmetry of minimizers. The main tool is the rearrangement (1.4) of a given minimizer to a family of radially symmetric functions, defined in Lemma 1.2.1. In contrast to this, the method used in [7] (see also [11] and [18]) is based on rearranging by averaging on concentric spheres. This, too, yields a radially symmetric function, however the disadvantage is that its minimizing property can only be shown for convex G (using Jensen's inequality), whereas our method does not require any conditions on G at all.

Theorem 1.2.4. *Assume (W_1), (W_{sym}) and (G_0). Furthermore assume that Ω is a ball centered at $0 \in \mathbb{R}^N$, W is nondecreasing in radial direction (i.e., $s \mapsto \tilde{W}(s)$ is nondecreasing for $s \geq 0$) and that E_0 has a global minimizer u in $W^{1,p}_0$. Then the following holds:*

(i) At least one global minimizer of E_0 is radially symmetric.

(ii) Any minimizer u which satisfies

$$\tilde{W}(\nu) > \tilde{W}(\partial_r u(x)) \textit{ whenever } |\nu| > |\partial_r u(x)|, \textit{ for a. e. } x, \tag{1.9}$$

is radially symmetric. Here, $\partial_r u(x) := x \cdot \nabla u(x)/|x|$ denotes the partial derivative of u in radial direction.

(iii) *Suppose that (G_1) holds. Then every minimizer u of E_0 satisfies (1.9) and thus is radially symmetric. Furthermore, u is either nonnegative or nonpositive in $B_R(0)$. Here, the latter case can occur only if $G(u) \equiv G(-u)$, so that $|u| = -u$ is a minimizer, too, then. If u is nonnegative and $\tilde{W}$ is constant on $[-M_0, M_0]$ for a $M_0 \geq 0$ (note that $M_0 = 0$ is allowed), then we have $\partial_r u \leq -M_0$ almost everywhere; in particular, u is decreasing in radial direction.*

Proof. **(i) Radial symmetry of one minimizer:**
In order to show radial symmetry of a minimizer u, we first consider the family $u_\theta \in W_0^{1,p}(\Omega)$, $\theta \in S^{N-1}$, of radially symmetric functions, defined as in Lemma 1.2.1. They satisfy

$$\int_{S^{N-1}} E_0(u_\theta) d\theta \leq \mathrm{Vol}_{N-1}(S^{N-1}) E_0(u). \tag{1.10}$$

This can be observed in the following way: The function W is radially symmetric by (W_{sym}) and nondecreasing in radial direction, so that (1.6) implies that

$$W(\nabla u_\theta(r\theta)) \leq W(\nabla u(r\theta)) \tag{1.11}$$

for almost every $r \in (0, R)$ and $\theta \in S^{N-1}$. Consequently,

$$\begin{aligned}
&\int_{S^{N-1}} E_0(u_\theta) d\theta \\
&= \int_{S^{N-1}} \int_{S^{N-1}} \int_0^R \left[W(\nabla u_\theta(r\psi)) + G(u_\theta(r\psi))\right] r^{N-1} dr d\psi d\theta \\
&= \int_{S^{N-1}} \int_{S^{N-1}} \int_0^R \left[W(\nabla u_\theta(r\theta)) + G(u_\theta(r\theta))\right] r^{N-1} dr d\psi d\theta \\
&\qquad \text{since } u_\theta \text{ is radially symmetric and } W \text{ satisfies } (W_{sym}) \\
&\leq \int_{S^{N-1}} \int_{S^{N-1}} \int_0^R \left[W(\nabla u(r\theta)) + G(u(r\theta))\right] r^{N-1} dr d\psi d\theta \\
&\qquad \text{due to (1.11)} \\
&= \mathrm{Vol}_{N-1}(S^{N-1}) E_0(u).
\end{aligned}$$

Since u is a minimizer, we know that $E_0(u) \leq E_0(u_\theta)$ for a. e. $\theta \in S^{N-1}$ (every θ such that $u_\theta \in W_0^{1,p}$). The only way this can coincide with (1.10) is if

$$E_0(u) = E_0(u_\theta), \quad \text{for a. e. } \theta \in S^{N-1}, \tag{1.12}$$

i.e., the radially symmetric function u_θ is a minimizer, too, for almost every θ.

(ii) Radial symmetry of all minimizers satisfying (1.9):
First observe that as a consequence of the calculation in (i), (1.12) is possible

only if equality holds in (1.11), for a. e. r and θ. By virtue of (1.9) and (1.6), this implies that

$$|\nabla u_\theta(r\theta)| = |\partial_r u(r\theta)| = |\nabla u(r\theta)|, \quad \text{for a. e. } r,\ \theta.$$

Thus the vector field $\nabla u(x)$ is colinear to x almost everywhere in $B_R(0)$. Since the only gradient fields on $B_R(0)$ with such a property are gradients of radially symmetric potentials, as seen in Lemma 1.2.3, this proves the radial symmetry of u.

(iii) Radial symmetry and common properties of all minimizers, assuming (G_1):

We define a rearrangement v_θ of the radially symmetric minimizers u_θ by setting

$$v_\theta'(r) := -\max\left\{\nu \geq 0 \mid \tilde{W}(\nu) = \tilde{W}(|u_\theta'(r)|)\right\} \text{ and } v_\theta(r) := -\int_r^R v_\theta'(s)ds.$$

Since $\tilde{W}$ is an even function by (W_{sym}),

$$\tilde{W}(v_\theta'(r)) = \tilde{W}(u_\theta'(r)) \text{ for every } r \in (0, R). \tag{1.13}$$

On the other hand, by the monotonicity of G assumed in (G_1),

$$G(v_\theta(r)) \leq G(u_\theta(r)) \text{ for every } r \in (0, R), \tag{1.14}$$

because obviously $|v_\theta| \geq |u_\theta|$. Now, (1.13) and (1.14) imply that

$$\begin{aligned} E_0(v_\theta) &= \mathrm{Vol}_{N-1}(S^{N-1}) \int_0^R \left[\tilde{W}(v_\theta') + G(v_\theta)\right] r^{N-1} dr \\ &\leq \mathrm{Vol}_{N-1}(S^{N-1}) \int_0^R \left[\tilde{W}(u_\theta') + G(u_\theta)\right] r^{N-1} dr = E_0(u_\theta). \end{aligned} \tag{1.15}$$

Remembering that u_θ is a global minimizer for E_0, we conclude that equality holds in (1.15) and thus also in (1.14), for every r, i.e.,

$$G(v_\theta) = G(u_\theta) \text{ on } (0, R), \tag{1.16}$$

By virtue of (G_1), the latter entails that $|v_\theta| = |u_\theta|$ and, consequently, $|v_\theta'| = |u_\theta'|$ almost everywhere. Since v_θ is decreasing, this implies that u_θ' cannot change sign on $(0, R)$, and thus

$$\text{either } u_\theta \equiv v_\theta \text{ or } u_\theta \equiv -v_\theta. \tag{1.17}$$

Furthermore, by the definition of v_θ' and the monotonicity of $\tilde{W}$, we have that

$$\tilde{W}(u_\theta'(r)) < \tilde{W}(\nu) \text{ whenever } |\nu| > |u_\theta'|,$$

for almost every $r \in (0, R)$. Thus (1.9) holds, and (ii) yields that $u = u_\theta$ for almost every θ; in particular, u is radially symmetric. The remaining properties of u claimed in the theorem now follow directly from (1.17), (1.16) and the definition of the v_θ. □

Remark 1.2.5. In particular, in the third part of Theorem 1.2.4 we have shown that the radial derivative of a minimizer does not take values on an interval where $\tilde{W}$ is constant. Actually, this result can be extended to parts where $\tilde{W}$ (respectively, $\tilde{W}^c$) is affine (but not necessarily constant), cf. Proposition 1.3.4 in Section 1.3. In the special case where G is convex, this was observed in [7] (see [8] for the proof).

Concluding this section, we derive a condition for the radial derivative of a bounded radially symmetric minimizer at the origin, which can be interpreted as a replacement for the second Weierstrass–Erdmann corner condition at this point. Note that we can assume that minimizers are bounded (belong to L^∞). Actually, all global minimizers are known to satisfy this, even in a more general setting not restricted to radial symmetry, cf. [30]. In the radial symmetric case, every critical point belongs to $W^{1,\infty}$ as we show in Proposition 1.5.3 below.

Proposition 1.2.6. *Assume that $\Omega = B_R(0)$ is a ball centered at the origin and that (W_0), (W_1), (W_{sym}), (G_0) and (G_1) are satisfied. Furthermore assume that $\tilde{W}$ is nondecreasing and that G is locally Lipschitz continuous. Then any radially symmetric and radially decreasing minimizer $u \in W_0^{1,p}(\Omega) \cap L^\infty(\Omega)$ of E_0 satisfies*

$$\lim_{r \to 0} \partial_r u(r) \to -M, \tag{1.18}$$

where M is given by (1.1), for a suitable representative of the L^p-function $\partial_r u(r)$.

Proof. Fix $\epsilon > 0$. For each $\delta \in (0, R)$ consider the set

$$I_\epsilon^\delta := \{ r \in (0, \delta) \mid \partial_r u(r) \leq -M - \epsilon \}.$$

We show that for each $\epsilon > 0$, there is a corresponding $\delta > 0$ such that I_ϵ^δ is of zero measure, entailing (1.18); remember that $u' \leq -M$ on $(0, R)$ by Theorem 1.3.1 (iii). For this purpose we define a radially symmetric function $u_\delta : B_R(0) \to \mathbb{R}$ such that in radial coordinates

$$\partial_r u_\delta(r) := \begin{cases} -M & \text{if } r \in I_\epsilon^\delta \\ \partial_r u(r) & \text{if } r \in (0, R) \setminus I_\epsilon^\delta, \end{cases} \quad \text{and } u_\delta(r) := -\int_r^R \partial_r u_\delta(s)\, ds.$$

Observe that $0 \leq u_\delta \leq u$ and $u_\delta \in W_0^{1,p}(B_R(0))$ for each δ. To compare the energies of u and u_δ, first note that for fixed ϵ, there exists a constant $c_\epsilon > 0$ such that

$$\tilde{W}(M) - \tilde{W}(\xi) \leq -c_\epsilon \left| -M - \xi \right|, \text{ whenever } \xi \leq -M - \epsilon \tag{1.19}$$

since $\tilde{W}$ is coercive by (W_1) and $W(\xi) > W(-M)$ whenever $\xi < -M$. Furthermore, we have that

$$\left| G(\mu_1) - G(\mu_2) \right| \leq L \left| \mu_1 - \mu_2 \right|, \tag{1.20}$$

for every pair $\mu_1, \mu_2 \in [-\|u\|_\infty, \|u\|_\infty]$ with a constant $L \geq 0$ independent of μ_1 and μ_2, because G is locally Lipschitz continuous. Now we estimate the energy difference as follows:

$$\begin{aligned}
&(E_0(u_\delta) - E_0(u)) \operatorname{Vol}_{N-1}(S^{N-1})^{-1} \\
&= \int_{I_\epsilon^\delta} \left[\tilde{W}(u_\delta') - \tilde{W}(u')\right] r^{N-1}\, dr + \int_0^\delta [G(u_\delta) - G(u)]\, r^{N-1}\, dr \\
&\leq -c_\epsilon \int_{I_\epsilon^\delta} |u_\delta' - u'|\, r^{N-1}\, dr + \int_0^\delta L\, |u_\delta - u|\, r^{N-1}\, dr \\
&\qquad \text{due to (1.19) and (1.20)} \\
&\leq -c_\epsilon \int_{I_\epsilon^\delta} |u_\delta' - u'|\, r^{N-1}\, dr + \int_0^\delta L \left(\int_r^\delta |u_\delta'(s) - u'(s)|\, s^{N-1}\, ds \right) dr \\
&\leq \int_{I_\epsilon^\delta} |u_\delta' - u'|\, r^{N-1}\, dr\, (-c_\epsilon + \delta L)
\end{aligned}$$

Since the second factor converges to $-c_\epsilon < 0$ as $\delta \to 0$, the whole expression eventually becomes negative unless

$$0 = \int_{I_\epsilon^\delta} |u_\delta' - u'|\, r^{N-1}\, dr \geq \int_{I_\epsilon^\delta} \epsilon r^{N-1}\, dr$$

for small δ, which entails that I_ϵ^δ is of measure zero. □

1.3 Existence and properties of minimizers for nonconvex potentials

For nonconvex W, the existence of minimizers of E_0 heavily depends on the shape of W, G and the boundary conditions imposed on u.

Without assuming radial symmetry, the existence of a minimizer of E_0 in $W_0^{1,p}$ is known if G does not have strict local minima and (roughly speaking) does not oscillate too fast, cf. [4], provided that the convex hull W^c of W is affine on every connected subset of $\mathbb{R}^N$ on which it differs from W. However, this behavior of the convex hull is by no means typical – usually W^c will be affine only along suitable one-dimensional lines wherever it differs from W. In such a case, the existence of a minimizer cannot be guaranteed in general, as shown in the next section. Also, if G is convex and coercive, or constant, then the infimum subject to Dirichlet boundary conditions typically is not attained. For intermediate behavior of G, the answer to the question of existence of minimizers is unknown.

More can be said in the radially symmetric case which has been studied in [7, 8, 11, 18]. There, the existence of radially symmetric minimizers is obtained assuming that G is convex and strictly decreasing (or strictly increasing). In the following, we show that one can drop convexity of G if G is of class C^2. In particular, this encompasses our prototype $G(\mu) = -\mu^2$. The cornerstone for this result has already been laid in Theorem 1.2.4.

Theorem 1.3.1. *Assume* (W_0)*–*(W_2)*,* (W_{sym}) *and* (G_0)*–*(G_2)*, and that* Ω *is a ball in* $\mathbb{R}^N$ *centered at the origin. In addition, suppose that* G *is either convex, strictly concave, or of class* C^2*. Then there exists a global minimizer of* E_0 *in* $W_0^{1,p}(\Omega)$*. Furthermore, every minimizer* u *of* E_0 *is radially symmetric and satisfies* $\tilde{W}(|\nabla u|) = \tilde{W}^c(|\nabla u|)$ *a. e., where* $\tilde{W}^c$ *is the convex hull of* $\tilde{W}$ *defined in (1.2). Moreover,* $|u|$ *always is a minimizer, too, and* $|u|$ *is radially decreasing; more specific, its derivative in radial direction satisfies* $\partial_r |u| \leq -M$ *almost everywhere and*

$$\partial_r |u| (x) \to -M \text{ as } |x| \to 0, \tag{1.21}$$

where M *is defined in (1.1).*

Proof. The proof of existence is based on relaxation . First we consider the relaxed Potential

$$E_0^c(u) := \int_{B_R(0)} W^c(\nabla u) + G(u)\, dx,$$

and note that W^c is convex, continuous and satisfies the same coercivity condition (W_1) as W. The functional E_0^c has a minimizer: Any minimizing sequence for E_0 in $W_0^{1,p}(\Omega)$ is bounded in this space by the coercivity of E_0^c inherited from E_0. Thus it converges weakly up to a subsequence, and the weak limit $u \in W_0^{1,p}(\Omega)$ is a minimizer due to the weak lower semicontinuity of E_0^c, cf. [12]. As a consequence of Theorem 1.2.4 applied to E_0^c, u is radially symmetric, nonnegative and decreasing in radial direction. We now have to show that

$$W(\nabla u) = W^c(\nabla u) \text{ almost everywhere} \tag{1.22}$$

The convex hull $\tilde{W}^c$ is affine on every connected component of the detachment set $\{t \mid \tilde{W}(t) > \tilde{W}^c(t)\}$. Note that the components are open since $\tilde{W}^c$ is continuous, and each one is bounded due to (W_1). Since ∇u does not have support on the constant parts of $\tilde{W}^c$ due to Theorem 1.2.4 (iii), it is enough to consider all components $H \subset (0, \infty)$ of the detachment set such that $\tilde{W}^c$ is affine and strictly increasing on H. There are at most countably many of those components, and thus it suffices to show that $S := \{r \in (0, R) \mid \partial_r u(r) \in H\}$ is of measure zero, for each such H. If G is convex, the set S is of measure zero as shown in [8] (if, in addition, G is of class C^2, one can use Proposition 1.3.4 instead). If G is strictly concave or if G is of class C^2, we arrive at the same conclusion by virtue of Proposition 1.3.5 below.

Since as a consequence of the Relaxation Theorem (see for example [12], Chapter 5) the infima of E_0^c and E_0 coincide, (1.22) immediately entails that u is a minimizer of E_0, too, concluding the proof of existence. The claimed qualitative properties of an arbitrary minimizer can be deduced as follows: Again by the Relaxation Theorem and the obvious fact that $E_0(u) \geq E_0^c(u)$, any minimizer u of E_0 is a minimizer of the relaxed potential E_0^c, too. Theorem 1.2.4 (iii) now yields the assertion with the exception of (1.21). The latter is a consequence of Proposition 1.2.6 applied to E_0^c. □

We now derive two results which in particular rule out the possibility that a radially symmetric and radially decreasing local minimizer has support on an interval where $\tilde{W}$ is affine, thereby providing the missing piece in the proof of Theorem 1.3.1 above. We need a few measure-theoretic notions:

Definition 1.3.2 (Lebesgue points and points of density). Let $f : \mathbb{R} \to \mathbb{R}$ be locally integrable and let $S \subset \mathbb{R}$ be Lebesgue-measurable. We call $s \in \mathbb{R}$ a *Lebesgue-point* of f if and only if

$$\frac{1}{h} \int_0^h |f(s+t) - f(s)| \, dt \to 0 \text{ as } h \to 0 \ (h \in \mathbb{R}).$$

Furthermore, we call $s \in \mathbb{R}$ a (measure-theoretic) *point of density* of S if and only if

$$\lim_{\delta \to 0} \frac{\mathrm{Vol}_1 \left(S \cap (s-\delta, s+\delta)\right)}{2\delta} = 1.$$

Remark 1.3.3. Almost all points of $\mathbb{R}$ are Lebesgue points of f, for an arbitrary function $f \in L^1_{\mathrm{loc}}(\mathbb{R})$. Almost all points of a measurable set $S \subset \mathbb{R}$ are points of density of S. In particular, if the set of points of density of S in $\mathbb{R}$ is of measure zero, then so is S. Furthermore, each point of density of S is an accumulation point of other points of density. For a proof of the first two assertion see for example [16]. The latter two are immediate consequences.

The proposition below is a variant of a result of A. Cellina and S. Perrotta [7, 8]:

Proposition 1.3.4. *Assume that $\Omega = B_R(0)$ is a ball, that W satisfies (W_0)–(W_2) and (W_{sym}) and that G is of class C^2. Furthermore suppose that $\tilde{W}$ is affine and increasing on an open interval $H \subset (0, \infty)$, i.e.*

$$\tilde{W}(t) = \alpha t + \beta \text{ for every } t \in H, \tag{1.23}$$

where $\alpha > 0$ and $\beta \in \mathbb{R}$ are constants. Let $u \in W_0^{1,p}(B_R(0))$ be a local extremal of E_0 which is radially symmetric and satisfies $\partial_r u \le 0$ on $B_R(0)$. Moreover let $r_0 \in (0,R)$ be a Lebesgue point of u' as well as a point of density of the set

$$S := \{r \in (0,R) \mid -\partial_r u(r) \in H\}.$$

Then we have that

$$G''(u(r_0)) \le -\alpha \frac{N-1}{\sup H} \cdot \frac{1}{r_0^2} < 0.$$

Moreover, the assertion above also holds if "local extremal" is replaced by "critical point" as long as E_0 is Gâteaux-differentiable at u.

Proof. Assume that u is a local extremal, for example (w.l.o.g.) a local minimizer. Our first aim is to derive the strong Euler–Lagrange equation (1.25) below, which is an immediate consequence of the Fundamental Lemma of Du Bois–Reymond if E_0 is differentiable at u and u is a critical point. We consider radially symmetric test functions $\varphi \in W^{1,\infty}(B_R(0))$ such that φ has compact support in $(0,R)$ and the following holds for all $r \in (0,R)$:

$$-u'(r) - h\varphi'(r) \in H \text{ for every } h \in [-1,1], \text{ wherever } \varphi'(r) \neq 0.$$

In particular, the latter implies that $\varphi' = 0$ outside of S (choose $h = 0$). An example for a test function satisfying these properties is constructed below. For every such φ whose norm in $W^{1,\infty}$ is sufficiently small,

$$\begin{aligned} 0 &\le \frac{1}{\mathrm{Vol}_{N-1} S^{N-1}} \left[E_0(u+\varphi) - E_u(u)\right] \\ &= \int_0^R \left[-\alpha\varphi' + G(u+\varphi) - G(u)\right] r^{N-1} dr \\ &= \int_0^R \left[\frac{N-1}{r}\alpha\varphi + G(u+\varphi) - G(u)\right] r^{N-1} dr, \end{aligned}$$

due to (1.23) and integration by parts. Since G is of class C^1, this entails

$$0 = \int_0^R \left[\frac{N-1}{r}\alpha + G'(u)\right] r^{N-1}\varphi \, dr. \tag{1.24}$$

Moreover, we infer that

$$\frac{N-1}{r_0}\alpha + G'(u(r_0)) = 0 \text{ whenever } r_0 \in (0,R) \text{ is a point of density of } S \tag{1.25}$$

by constructing a suitable admissible test function to rule out the alternative: Assume (w.l.o.g.) that $\frac{N-1}{r_0}\alpha + G'(u(r_0)) > 0$ at a point of density $r_0 \in (0,R)$ of S. Thus, by continuity,

$$\frac{N-1}{r}\alpha + G'(u(r)) > 0, \text{ for every } r \text{ in a vicinity } (a_1,a_2) \text{ of } r_0. \tag{1.26}$$

For arbitrary $b \in (a_1,a_2)$ we define $\varphi_b(r) := -\int_r^R \varphi_b'(t)\,dt$ as follows:

$$\varphi_b' := \begin{cases} \frac{1}{2}\operatorname{dist}(|u'|\,;\partial H) & \text{on } (a_1,b)\cap S, \\ -\frac{1}{2}\operatorname{dist}(|u'|\,;\partial H) & \text{on } (b,a_2)\cap S, \\ 0 & \text{elsewhere.} \end{cases}$$

By continuity, there is a point $b_0 \in (a_1,a_2)$ such that $\varphi_{b_0}(a_1) = 0$. Thus $\varphi_{b_0} \geq 0$ on $(0,R)$ and $\operatorname{supp}\varphi_{b_0} \subset [a_1,a_2] \subset (0,R)$. Hence φ_{b_0} is admissible as a test function for (1.24), contradicting (1.26). Here, note that φ_{b_0} does not vanish almost everywhere since $(a_1,a_2)\cap S$ is of positive measure – remember that $r_0 \in (a_1,a_2)$ is a point of density of S.

Now fix a point $r_0 \in (0,R)$ of density of S which also is a Lebesgue point of u'. Since points of density are never isolated, there exists a sequence $h_n \neq 0$, $h_n \to 0$ such that $r_0 + h_n$ is a point of density of S, too, for every n. Subtracting the equations (1.25) at $r_0 + h_n$ and r_0 and dividing by h_n, we get

$$(N-1)\frac{\alpha}{h_n}\left(\frac{1}{r_0+h_n} - \frac{1}{r_0}\right) + \frac{1}{h_n}\left[G'(u(r_0+h_n)) - G'(u(r_0))\right] = 0. \tag{1.27}$$

for every $n \in \mathbb{N}$. By the mean value theorem,

$$\frac{G'(u(r_0+h_n)) - G'(u(r_0))}{h_n} = G''(U_n)\frac{u(r_0+h_n)-u(r_0)}{h_n}$$

where U_n lies between $u(r_0+h_n)$ and $u(r_0)$. We also have that

$$0 > \frac{u(r_0+h_n)-u(r_0)}{h_n} = \frac{1}{h_n}\int_0^{h_n} u'(r_0+t)\,dt =: -d_n,$$

where $u(r_0+h_n) = u(r_0)$ is impossible since this would contradict (1.27). Consequently, (1.27) implies that

$$G''(U_n) = -\alpha\frac{N-1}{d_n}\frac{1}{h_n}\left(-\frac{1}{r_0+h_n} + \frac{1}{r_0}\right). \tag{1.28}$$

Since r_0 is a Lebesgue point of u' and a point of density of $S = \{-u' \in H\}$, we infer that $-u'(r_0) \in \overline{H}$ and

$$0 \leq \lim_{n\to\infty} d_n = -u'(r_0) \leq \sup H.$$

Furthermore,

$$G''(U_n) \to G''(u(r_0)) \quad \text{and} \quad \frac{1}{h_n}\left(-\frac{1}{r_0+h_n}+\frac{1}{r_0}\right) \to \frac{1}{r_0^2},$$

as $n \to \infty$; remember that $h_n \to 0$. Thus passing to the limit in (1.28) yields the assertion. □

In particular, for a local minimizer u, we have shown that G is strictly concave near $u(r_0)$ whenever r_0 is a point of density of the set where $\partial_r u$ lies in an affine part of $\tilde{W}$ (as well as a Lebesgue point of $\partial_r u$). In fact, this contradicts the stability of u:

Proposition 1.3.5. *Assume that $\Omega = B_R(0)$ is a ball, that W satisfies (W_0)-(W_2) and (W_{sym}). Furthermore suppose that $\tilde{W}$ is affine and increasing on an open interval $H \subset (0, \infty)$, i.e.*

$$\tilde{W}(t) = \alpha t + \beta \text{ for every } t \in H, \tag{1.29}$$

where $\alpha > 0$ and $\beta \in \mathbb{R}$ are constants. Let $u \in W_0^{1,p}(B_R(0))$ be a radially symmetric local minimizer of E_0. Then any point $r_0 \in (0, R)$ such that

$$G \text{ is strictly concave in a vicinity of } u(r_0)$$

is not a point of density of $S := \{r \in (0, R) \mid -\partial_r u(r) \in H\}$. In particular, if G is of class C^2 and $\partial_r u \leq 0$ on $B_R(0)$ then S is of measure zero due to Proposition 1.3.4 and Remark 1.3.3.

Proof. The proof is indirect. Assume that $r_0 \in (0, R)$ is a point of density of S. We choose $\delta > 0$ and a vicinity (a_1, a_2) of r_0, $0 < a_1 < a_2 < R$, small enough such that

$$\begin{aligned} &G \text{ is strictly concave on } (-2\delta + u(r_0), 2\delta + u(r_0)) \text{ and} \\ &|u(r) - u(r_0)| < \delta \text{ whenever } r \in (a_1, a_2). \end{aligned}$$

Now define a radially symmetric test function $\varphi \in W^{1,\infty}(B_R(0))$ such that the support of φ is a compact subset of $(0, R)$,

$$-u'(r) - h\varphi'(r) \in H \text{ for every } h \in [-1, 1], \text{ wherever } \varphi'(r) \neq 0, \tag{1.30}$$

$$\varphi \neq 0 \text{ on a set of positive measure and} \tag{1.31}$$

$$\|\varphi\|_{L^\infty} < \delta. \tag{1.32}$$

Such a test function can be obtained analogously to the definition of φ_{b_0} in the proof of Proposition 1.3.4: For arbitrary $b \in (a_1, a_2)$ let $\varphi_b(s) := -\int_s^R \varphi_b'(t)\,dt$, where

$$\varphi_b' := \begin{cases} \frac{1}{2}\operatorname{dist}(|u'|\,;\partial H) & \text{on } (a_1, b) \cap S, \\ -\frac{1}{2}\operatorname{dist}(|u'|\,;\partial H) & \text{on } (b, a_2) \cap S, \\ 0 & \text{elsewhere,} \end{cases}$$

and choose $b_0 \in (a_1, a_2)$ in such a way that $\varphi_{b_0}(a_1) = 0$. Then the function $\varphi := \gamma\varphi_{b_0}$ fulfills our requirements, where $\gamma \in (0, 1]$ is a suitable scaling factor ensuring (1.32). Since u is a local minimizer of E_0, we have

$$\begin{aligned} 0 &\leq [E_0(u+\varphi) + E_0(u-\varphi) - 2E_0(u)]\,\frac{1}{\operatorname{Vol}_{N-1}(S^{N-1})} \\ &= \int_{a_1}^{a_2} [G(u+\varphi) + G(u-\varphi) - 2G(u)]\,r^{N-1}dr, \end{aligned} \tag{1.33}$$

due to (1.30) and (1.29), at least as long as γ (and thus $\|\varphi\|_{W^{1,p}}$) is small enough. However G is strictly concave on an interval containing all possible values of its arguments in (1.33), and thus $G(u+\varphi) + G(u-\varphi) - 2G(u) < 0$ wherever $\varphi \neq 0$, which contradicts (1.33) by virtue of (1.31). □

1.4 A nonexistence result in a setting without radial symmetry

Existence of a minimizer is no longer clear if W is not radially symmetric. For $G = 0$, conditions for nonattainment have been obtained in [27, 9, 17]. In the example below, we observe that in case of a double well, the infimum of E_0 is not necessarily attained even for our prototype $G(u) = -u^2$. This shows a fundamental difference to the one-dimensional case where (essentially) concavity of G is sufficient to assure existence of a minimizer (see [6]).

Example 1.4.1. We consider the functional

$$\tilde{E}_0(u) := \int_\Omega [\tilde{W}(\nabla u) + \alpha\tilde{G}(u)]\,dx, \;\; u \in W_0^{1,4}(\Omega),$$

on $\Omega := (-1, 1)^2 \subset \mathbb{R}^2$, where $\alpha > 0$ is a parameter,

$$\begin{aligned} &\tilde{W}(x, y) := (x^2 - 1)^2 + y^4 + y^2, \text{ and} \\ &\tilde{G}(\mu) := -\mu^2 \quad \text{or} \quad \tilde{G}(\mu) := -\mu\,|\mu|\,. \end{aligned}$$

In the second case, $\tilde{G}$ is strictly monotonic. We claim that $\tilde{E}_0$ does not have a minimizer whenever

$$\alpha < \alpha_0 := \inf \left\{ \int_{-1}^{1} (v')^2 \, dy \mid v \in W_0^{1,4}(-1,1) \text{ with } \|v\|_{L^2(-1,1)} = 1 \right\}.$$

Note that $\alpha_0 > 0$; actually $\alpha_0 = \pi^2/4$, which is the first eigenvalue of $-\frac{\partial^2}{\partial y^2}$ on $(-1,1)$ subject to Dirichlet boundary conditions. For our argument, the case of $\tilde{G}(\mu) = -\mu\,|\mu|$ reduces to that of $\tilde{G}(\mu) = -\mu^2$: Any minimizer u of E_0 in the case $\tilde{G}(\mu) = \mu\,|\mu|$ is nonnegative since otherwise $|u|$ would have less energy. Thus u is a minimizer of E_0 with $\tilde{G}(\mu) = -\mu^2$, too. If $\alpha \leq \alpha_0$, any function $u \in W_0^{1,4}((-1,1)^2)$ satisfies

$$\int_{-1}^{1} [u_y(x,y)^4 + u_y(x,y)^2 - \alpha u(x,y)^2] \, dy \geq 0 \tag{1.34}$$

for almost every $x \in (-1,1)$. Note also that for a fixed x, equality in (1.34) implies that $u(x,\cdot) \equiv 0$. Consequently,

$$\begin{aligned} \tilde{E}_0(u) &= \int_{-1}^{1}\int_{-1}^{1} [(u_x^2 - 1)^2 + u_y^4 + u_y^2 - \alpha u^2] \, dy \, dx \\ &\geq \int_{-1}^{1}\int_{-1}^{1} (u_x^2 - 1)^2 \, dy \, dx \geq 0, \end{aligned}$$

and $\tilde{E}_0(u) = 0$ is impossible since this would imply that on the one hand, equality holds in (1.34) for a. e. x, entailing $u \equiv 0$, and on the other hand, $(u_x^2 - 1)^2 = 0$ almost everywhere. However it is not difficult to find a sequence (u_n) such that $\tilde{E}_0(u_n) \to 0$; simply construct $u_n \in W_0^{1,4}((-1,1)^2)$ in such a way that

$$\begin{gathered} u_n \leq \frac{1}{n} \text{ on } (-1,1)^2, \quad |\nabla u_n| \leq 1 \text{ on } (-1,1)^2, \quad \text{and} \\ \mathrm{Vol}_2\{(x,y) \in (-1,1)^2 \mid \frac{\partial}{\partial x} u_n(x,y) \neq \pm 1\} \leq \frac{1}{n}. \end{gathered}$$

Thus the infimum of $\tilde{E}_0$ is 0, and it is not attained.

Although the functions $\tilde{W}$ and $\tilde{G}$ used in the example are of special form, the idea can be easily extended to a nonexistence result for a rather general class potentials with multiple wells in a single direction:

Theorem 1.4.2. *Consider the functional*

$$E_0(u) := \int_{\Omega} [W(\nabla u) + \alpha G(u)] \, dx, \; u \in W_0^{1,2}(\Omega),$$

on a smooth bounded domain $\Omega \subset \mathbb{R}^N$. *Here,* $\alpha \in \mathbb{R}$ *is a parameter and* $W : \mathbb{R} \times \mathbb{R}^{N-1} \to \mathbb{R}$ *and* $G : \mathbb{R} \to \mathbb{R}$ *are measurable functions satisfying*

$$W(\xi, \eta) \geq W_1(\xi) + |\eta|^2, \; W(m_1, 0) = W(m_2, 0) = 0, \tag{1.35}$$

$$G(\mu) \geq -|\mu|^2, \; \text{and } G(0) = 0. \tag{1.36}$$

for every $\xi \in \mathbb{R}$, $\eta \in \mathbb{R}^{N-1}$ *and* $\mu \in \mathbb{R}$. *Here,* $m_1 < 0 < m_2$ *are constants and* $W_1 : \mathbb{R} \to \mathbb{R}$ *is a measurable function such that*

$$W_1(x) \geq 0 \text{ for every } x \in \mathbb{R}, \; W_1(m_1) = W_1(m_2) = 0 \text{ and } W_1(0) > 0.$$

Then E_0 *does not attain its minimum whenever*

$$\alpha < \alpha_0 := \inf \left\{ \frac{\int_{\Omega_x} |\nabla v(y)|^2 \, dy}{\int_{\Omega_x} v(y)^2 \, dy} \;\middle|\; x \in P_1\Omega \text{ and } v \in W_0^{1,2}(\Omega_x) \setminus \{0\} \right\},$$

where $\Omega_x := \Omega \cap \big(\{x\} \times \mathbb{R}^{N-1}\big)$ *and* $P_1\Omega := \{x \in \mathbb{R} \mid \Omega_x \neq \emptyset\}$.

Remark 1.4.3. (i) Using the well known fact that the optimal constant in Poincaré's inequality is attained on the ball, we can estimate α_0 from below by the first eigenvalue of the negative Laplacian subject to Dirichlet boundary conditions on a ball in $\mathbb{R}^{N-1}$ with volume $\sup\{\mathrm{Vol}_{N-1}\,\Omega_x \mid x \in P_1\Omega\}$. In particular, $\alpha_0 > 0$.

(ii) The missing constant in front of the term $|\eta|^2$ in (1.35) is not an restriction since one can get the general case by rescaling E_0 with an appropriate factor. Also, if the minima of W lie on an arbitrary line through 0 (not necessarily the first coordinate axis), on both sides of 0, this can be reduced to the situation in (1.35) by rotation of the domain.

(iii) The quadratic terms $|\eta|^2$ and $|\mu|^2$ in (1.35) and (1.36), respectively, can be replaced by different expressions to cover an even more general setting, for example one can use $|\eta|^p$ and $|\mu|^p$ instead, with an arbitrary $p > 1$. The Raleigh quotient in the definition of α_0 as well as the space of admissible functions of course then have to be modified accordingly.

Proof. Since the proof relies on the same ideas already used in Example 1.4.1, we will only give a brief sketch. On the one hand, observe that the infimum of E_0 is less or equal to zero by constructing a sequence $u_n \in W_0^{1,p}(\Omega)$ such that

$$\begin{aligned}
&u_n \text{ is uniformly bounded in } W^{1,\infty}, \; u_n \to 0 \text{ in } L^\infty(\Omega), \text{and} \\
&\nabla u_n \in \{(m_1, 0), (m_2, 0)\} \text{ on } \Omega \setminus H_n, \\
&\text{where } H_{n+1} \subset H_n \text{ and } \mathrm{Vol}\, H_n \to 0 \text{ as } n \to \infty.
\end{aligned}$$

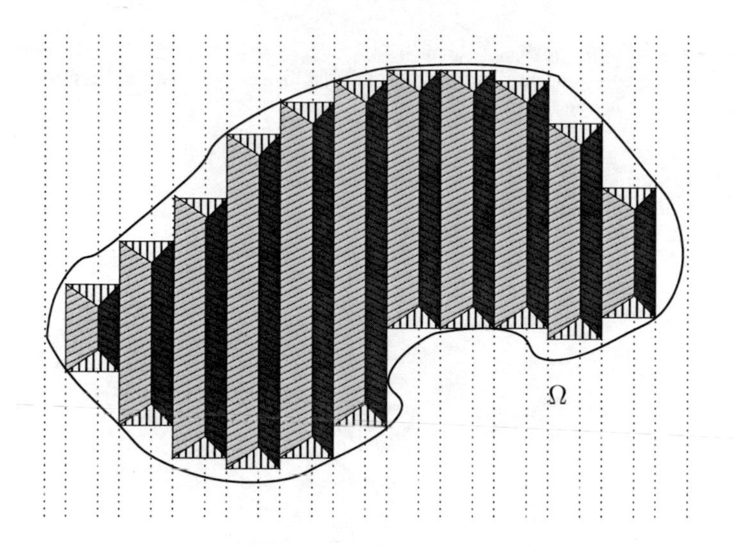

Figure 1.1: Construction of a minimizing sequence

(The easiest way to do this is by covering Ω with strips dividing Ω in direction of the first coordinate x as indicated in Figure 1.1. Then define a suitable jigsaw-like function which is affine on the strips with a gradient alternating between the two minima of W, except on a small area H_n near the boundary of Ω which is used to ensure that u_n satisfies the Dirichlet boundary conditions.) On the other hand, like in the example use the fact that

$$\int_{\Omega_x} [\nabla_y u(x,y)^2 - \alpha u(x,y)^2]\, dy \geq 0$$

for $\alpha < \alpha_0$ to conclude that $E_0 \geq 0$ where at the same time $E_0 \neq 0$ always. Thus the infimum 0 of E_0 is not a minimum. $\square$

1.5 Properties of critical points

In this section we consider radially symmetric solutions $u \in W_0^{1,p}(B_R(0))$ of the Euler–Lagrange equation corresponding to E_0:

$$\int_{B_R(0)} \nabla W(\nabla u) \cdot \nabla \varphi + G'(u)\varphi\, dx = 0 \qquad (EL_0)$$

for every radially symmetric test function $\varphi \in W_0^{1,p}(B_R(0))$.

Note that, assuming (W_{sym}), (W_0^{EL})-(W_2^{EL}) and (G_0^{EL})-(G_2^{EL}), E_0 is Fréchèt differentiable as a functional on $W_0^{1,p}(B_R(0))$, and its derivative in direction φ is given by the left hand side of (EL_0).

First, observe that due to the radial symmetry, a (weak) solution u of (EL_0) is a classical solution, up to a possible singularity at 0:

Proposition 1.5.1. *Assume (W_0^{EL})-(W_2^{EL}), (W_{sym}) and (G_0^{EL})-(G_2^{EL}), and let $u \in W_{r,0}^{1,p}(B_R(0))$ be a radially symmetric solution of (EL_0). Then $\tilde{W}'(u')$ is of class C^1 on $(0,R]$, $u(R)=0$ and*

$$\frac{d}{dr}\left[r^{N-1}\tilde{W}'(u')\right] = r^{N-1}G'(u) \text{ on } (0,R]. \tag{EL_0^s}$$

Furthermore, for any local extremal u of E_0 in $W_{r,0}^{1,p}(B_R(0))$ the second Weierstrass-Erdmann corner condition is satisfied:

$$\tilde{W}(u') - u'\tilde{W}'(u') \text{ is continuous on } (0,R]. \tag{1.37}$$

Proof. In radial coordinates, (EL_0) reads

$$\int_0^R \left[\tilde{W}'(u')\varphi' + G'(u)\varphi\right] r^{N-1}dr = 0. \tag{1.38}$$

Since $u \in W_{r,0}^{1,p}(B_R(0))$ we immediately infer that u is continuous on $(0,R]$, using the one-dimensional Sobolev imbedding theorem. The Fundamental Lemma of the Calculus of Variations entails that $r \mapsto r^{N-1}\tilde{W}'(u')$ is continuously differentiable on $(0,R]$, and thus the same holds for $\tilde{W}'(u')$. Consequently, u even satisfies the strong version of (1.38), i. e. (EL_0^s), on $(0,R]$. The second Weierstrass-Erdmann corner condition is obtained in the standard way: For arbitrary but fixed $r_0 \in (0,R)$ apply the Fundamental Lemma to the local extremal $t \mapsto (s_u(t), v_u(t)) := (t, u(t))$ $(t \in (r_0,R))$ of the parametric variational problem associated to (E_0):

$$\tilde{E}_0(s,v) := \int_{r_0}^R \left[\tilde{W}(v'(t)/s'(t)) + G(v(t))\right](s(t))^{N-1}s'(t)\,dt,$$

where $(s,v) \in Y := (s_u, v_u) + \left[C_D^1[r_0,R] \times W_0^{1,p}(r_0,R)\right]$. For more details see Section A.1 in the appendix. □

For nonnegative solutions u one can find restrictions on the range of u' as a consequence of (G_1^{EL}). We are interested in that because regularity of $\tilde{W}'(u')$ implies regularity of u' if $\tilde{W}'$ is invertible with sufficiently smooth inverse on the range of u'.

Proposition 1.5.2. *Assume* (W_0^{EL})*-*(W_2^{EL})*,* (W_{sym}) *and* (G_0^{EL})*-*(G_2^{EL})*, and let* $u \in W_{r,0}^{1,p}(B_R(0))$ *be a radially symmetric nonnegative solution of* (EL_0)*. Then*

$$\tilde{W}'(u') \leq 0 \text{ on } (0,R), \tag{1.39}$$

and the function $r \mapsto r^{N-1}\tilde{W}'(u'(r))$ *is decreasing. If* $G'(\mu) < 0$ *for every* $\mu > 0$ *and* $u \not\equiv 0$ *is nonnegative and decreasing on* $(0,R)$ *then we even have that*

$$\tilde{W}'(u') < 0 \text{ on } (0,R). \tag{1.40}$$

In particular, $u' < -M$ *on* $(0,R)$ *provided that* $\tilde{W}' \geq 0$ *on* $[-M,0]$*.*

Proof. The monotonicity of $r \mapsto r^{N-1}\tilde{W}'(u'(r))$ is a direct consequence of (EL_0^s), (G_1^{EL}) and the nonnegativity of u.

The proof of (1.39) is indirect. Assume that $\tilde{W}'(u'(r_0)) =: 2\rho > 0$ at a point $r_0 \in (0,R)$. Since $\tilde{W}'(u')$ is continuous by Proposition 1.5.1, we even have that

$$\tilde{W}'(u') \geq \rho > 0 \text{ in an Interval } (r_1, r_2) \text{ containing } r_0.$$

Now, construct a radially symmetric test function $\varphi \in C_0^\infty(B_R(0))$ in such a way that (using radial coordinates)

$$\varphi(r) = \begin{cases} 1 & \text{if } r \in [0, r_1], \\ 0 & \text{if } r \in [r_2, R], \end{cases}$$

and $\varphi'(r) \leq 0$ whenever $r \in (r_1, r_2)$. Using φ as a test function in (EL_0), we get

$$\begin{aligned} 0 &= \int_0^R \left[\tilde{W}'(u')\varphi' + G'(u)\varphi\right] r^{N-1} dr \\ &\leq \int_{r_1}^{r_2} \rho\varphi' r^{N-1} dr < 0, \end{aligned}$$

since $G'(u)\varphi \leq 0$. Obviously this is a contradiction.

Finally, assume that $G'(\mu) < 0$ for every $\mu > 0$ and u is nontrivial, nonnegative and decreasing on $(0,R)$. If there was a point $r_0 \in (0,R)$ such that $\tilde{W}'(u'(r_0)) = 0$, then $\tilde{W}'(u'(r)) = 0$ for every $r \in (0, r_0)$ due to (1.39) and the monotonicity of $r \mapsto r^{N-1}\tilde{W}'(u'(r))$. In this case (EL_0^s) would imply that $G'(u) = 0$ on $(0, r_0)$ and thus $u \equiv 0$, first on $(0, r_0)$ and then everywhere since u is nonnegative and decreasing. However this contradicts our assumption that u is nontrivial, and thus $\tilde{W}'(u'(r)) < 0$ on $(0,R)$ as claimed. □

Our next result shows that solutions necessarily are elements of $W^{1,\infty}$.

Proposition 1.5.3. *Assume* (W_0^{EL})*-*(W_2^{EL})*,* (W_{sym}^{EL}) *and* (G_0^{EL})*-*(G_2^{EL})*, and let* $u \in W_0^{1,p}(B_R(0))$ *be a radially symmetric solution of* (EL_0)*. Then we have the a–priori estimate*

$$\|u\|_{W^{1,\infty}(B_R(0))} \leq C(\|u\|_{L^{p-1}(B_R(0))} + 1) \tag{1.41}$$

with a constant $C \geq 0$ *independent of* u*.*

Proof. Since the class of test functions in (EL_0) contains $C_0^\infty(B_R(0))$, no restriction is made on the value of test functions at $r = 0$. This entails the natural "boundary" condition

$$r^{N-1}\tilde{W}'(u'(r)) \to 0 \text{ as } r \to 0.$$

Thus by integrating (EL_0^s) from t to r and letting $t \to 0$, we get

$$r^{N-1}\tilde{W}'(u'(r)) = \int_0^r G'(u(s))s^{N-1}ds. \tag{1.42}$$

In order to prove the bound for u in $W^{1,\infty}(B_R(0))$, first observe that (W_{sym}^{EL}) and (W_1^{EL}) together imply that

$$\left|\tilde{W}'(s)\right| \geq c\,|s|^{p-1} - C,$$

for every $s \in \mathbb{R}$, with constants $c > 0$ and $C \in \mathbb{R}$. Using this estimate in conjunction with (G_2^{EL}) in (1.42) yields

$$c\,|u'(r)|^{p-1} - C \leq \frac{1}{r^{N-1}} \int_0^r (\nu_3\,|u(s)|^{p-1} + C)s^{N-1}ds, \tag{1.43}$$

where we replaced $\tilde{p}$ with p in (G_2^{EL}) (which then obviously is a weaker condition). Inequality (1.43) can be used to obtain a uniform bound on u' via bootstrapping: As a starting point, note that the integral on the right hand side is bounded by $\tilde{C}_1\,\|u\|_{L^{p-1}(B_R(0))}^{p-1} + \tilde{C}_2$, with constants $\tilde{C}_1, \tilde{C}_2 \geq 0$ independent of r and u. Thus we have

$$|u'(r)| \leq (\|u\|_{L^{p-1}(B_R(0))} + 1)(C_1 r^{-\frac{N-1}{p-1}} + C_2).$$

For a step in the bootstrap argument we now assume that for almost every $r \in (0, R)$

$$|u'(r)| \leq (\|u\|_{L^{p-1}(B_R(0))} + 1)\left(C_1 r^\alpha + C_2\right), \tag{1.44}$$

with a number $\alpha \geq -(N-1)/(p-1)$. By integration (w. l. o. g. assuming that $\alpha \neq -1$, just use a slightly smaller α in this case),

$$|u(r)| \leq (\|u\|_{L^{p-1}(B_R(0))} + 1)(C_3 r^{\alpha+1} + C_4).$$

Using this to replace u in (1.43), we get

$$\begin{aligned}
&c\,|u'(r)|^{p-1} - C \\
&\leq (\|u\|_{L^{p-1}(B_R(0))} + 1)^{p-1} \frac{1}{r^{N-1}} \int_0^r (C_5 s^{(\alpha+1)(p-1)+N-1} + C_6 s^{N-1})\, ds \\
&\leq (\|u\|_{L^{p-1}(B_R(0))} + 1)^{p-1} (C_7 r^{(\alpha+1)(p-1)+1} + C_8 r),
\end{aligned}$$

and thus

$$|u'(r)| \leq (\|u\|_{L^{p-1}(B_R(0))} + 1) \left(C_9 r^{\alpha+1+\frac{1}{p-1}} + C_{10} \right),$$

which means that we have increased the exponent α in (1.44) by $1 + 1/(p-1)$. Consequently, repeating the process will eventually yield (1.44) with nonnegative α and thus a uniform bound for u'. Since u is subject to Dirichlet boundary conditions at $r = R$, this also implies that u is uniformly bounded on $(0, R)$. □

As a byproduct, the bound in $W^{1,\infty}$ unveils a relation of the radial derivative of a solution at the center 0 and the Maxwell points of W:

Corollary 1.5.4. *Assume* (W_0^{EL})*-*(W_2^{EL}), (W_{sym}^{EL}) *and* (G_0^{EL})*-*(G_2^{EL}), *and let* $u \in W_0^{1,p}(B_R(0))$ *be a radially symmetric solution of* (EL_0). *Then*

$$\lim_{|x|\to 0} \tilde{W}'(\partial_r u(x)) = 0. \tag{1.45}$$

If $G' < 0$ *on* $\mathbb{R}^+$, $\tilde{W}' \geq 0$ *on* $[-M, 0]$, $\tilde{W}' < 0$ *on* $(-\infty, -M)$ *and* $u \not\equiv 0$ *is nonnegative and decreasing on* $(0, R)$, *then we even have that*

$$\begin{aligned}
&\partial_r u(x) < -M \ \text{ on } \overline{B_R(0)} \setminus \{0\}, \ \text{ and} \\
&\partial_r u(x) \to -M \ \text{ as } |x| \to 0.
\end{aligned} \tag{1.46}$$

Here, $M \geq 0$ *denotes the Maxwell point of* $\tilde{W}$ *defined in (1.1).*

Proof. By Proposition 1.5.3 there is a constant $C > 0$ such that $|G'(u)| \leq C$ on $(0, R)$, and going back to the integrated version (1.42) of (EL_0^s) we conclude that

$$\left|\tilde{W}'(u'(r))\right| \leq C \frac{1}{r^{N-1}} \int_0^r s^{N-1} ds = \frac{C}{N} r \to 0,$$

as $r \to 0$. The second part of the assertion is an immediate consequence of (1.45) and Proposition 1.5.2. □

1.6 Uniqueness of critical points

The aim of this section is to prove uniqueness of critical points of E_0, within a class of functions which contains both global minimizers and the singular limits of Chapter 2. Our argument only works under additional assumptions besides (W_0^{EL})-(W_2^{EL}), (W_{sym}^{EL}) and (G_0^{EL})-(G_2^{EL}). In order to formulate those, we first introduce some notation: Let

$$H(z) := \tilde{W}''(z) - \frac{1}{z}\tilde{W}'(z),\ H(0) := 0, \tag{1.47}$$

and

$$\beta(\rho, z) := \frac{H(\rho z)}{H(z)},\ \beta(\rho, 0) := \rho, \tag{1.48}$$

where $z \in \mathbb{R} \setminus \{0\}$ and $\rho \geq 1$. Note that H is of class C^1 and β is continuous (in particular at $z = 0$) if $\tilde{W}$ is of class C^3. The definition of H is motivated by the fact that the Euler–Lagrange equation (EL_0^s) can be written as

$$\tilde{W}''(u')\left(u'' + \frac{N-1}{r}u'\right) - \frac{N-1}{r}H(u')u' = G'(u), \tag{1.49}$$

provided the participating functions are smooth enough to evaluate the total derivative on the left hand side of (EL_0^s).

Theorem 1.6.1. *Assume that W is of class C^3 and satisfies (W_1^{EL}), (W_2^{EL}), (W_{sym}^{EL}),*

$$\tilde{W}'' \geq c > 0 \ on\ (-\infty, -M] \tag{1.50}$$

where M is given by (1.1),

$$H(z) > 0 \ if\ z \leq -M, \tag{1.51}$$

$$\beta(\rho, z) \geq 1 \ whenever\ \rho > 1 \ and\ z \leq -M, \ and \tag{1.52}$$

$$\beta(\rho, z)\tilde{W}''(z) < \tilde{W}''(\rho z) \ for\ every\ z \leq -M \ and\ \rho > 1. \tag{1.53}$$

Moreover, suppose that G satisfies (G_0^{EL})-(G_2^{EL}) and

$$\mu \mapsto \frac{G'(\mu)}{\mu} \ is\ nondecreasing\ on\ (0, \infty). \tag{1.54}$$

Then there is at most one nontrivial weak solution $u \in W_0^{1,p}(B_R(0))$ of (EL_0) with the following properties:

(i) u is radially symmetric.

(ii) $\partial_r u \leq -M$ *almost everywhere.*

(iii) $\partial_r u(r) \to -M$ *as* $r \to 0$.

Remark 1.6.2. The assumptions (1.51)–(1.53) as well as (1.54) are of technical nature, we are not able to give necessary conditions for uniqueness. They provide a framework which allows us to generalize the uniqueness result of [23] to the radially symmetric case, essentially using the same approach. Assuming (1.51), we have the following equivalent formulations of (1.52) and (1.53), respectively:

$$z \mapsto \tilde{W}''(z) - \frac{\tilde{W}'(z)}{z} \text{ is nonincreasing on } (-\infty, -M] \text{ and}$$
$$z \mapsto \frac{\tilde{W}'(z)}{z\tilde{W}''(z)} \text{ is strictly decreasing on } (-\infty, -M].$$

Note that the prototype function $\tilde{W}(z) := Az^4 - Bz^2$ $(A, B > 0)$ satisfies (1.51)–(1.53). In this case,

$$H(z) = 8Az^2, \quad \beta(\rho, z) = \rho^2 \quad \text{and}$$
$$\beta(\rho, z)\tilde{W}''(z) = 12A\rho^2 z^2 - 2B\rho^2 < 12A\rho^2 z^2 - 2B = \tilde{W}''(\rho z) \text{ if } \rho > 1.$$

Proof of Theorem 1.6.1. The proof is indirect. Suppose there exist two solutions u and v of (EL_0) with the properties (i)–(iii) which are not identical. Observe that u and v are classical solutions: As a consequence of Proposition 1.5.1 we know that $\tilde{W}'(\partial_r u)$ is continuously differentiable on $(0, R)$. Because of (1.50) and (ii), $\tilde{W}'$ is invertible on the interval $(-\infty, -M]$ where $\partial_r u$ takes its values and thus u (or v likewise) is of class C^2 on $(0, R]$; in particular, u and v satisfy (1.49) in $(0, R)$ in the classical sense. Due to (iii), we also have that $u, v \in C^1([0, R])$. Some further properties of u (or v respectively) which immediately follow from our assumptions are

$$u > 0 \text{ on } [0, R) \text{ and} \tag{1.55}$$
$$\Delta u \leq 0 \text{ on } (0, R). \tag{1.56}$$

The former is due to (ii) (and the fact that u is nontrivial if $M = 0$), whereas the latter is a consequence of equation (1.49), (ii), (1.50), (1.51) and (G_1^{EL}).

The outline of the proof is as follows: First, we scale v with a factor ρ such that ρv touches u at a point r_0. Then we use equation (1.49) which both u and v satisfy and the maximum principle to infer that u and ρv must coincide, which only is possible if $\rho = 1$.

We know that $\partial_r u(R) \neq \partial_r v(R)$ since otherwise u and v would be identical, because $u(R) = v(R)$ and both solve (1.49) which is an ODE of second order.

W. l. o. g. we assume that $\partial_r u(R) < \partial_r v(R)$. Now scale v in such away that it touches u from above, using the scaling factor

$$\rho := \inf \{s > 0 \mid sv \geq u \text{ on } (0, R)\} \in \mathbb{R}. \tag{1.57}$$

By construction we have that $\rho > 1$,

$$\rho v \geq u \text{ on } (0, R), \tag{1.58}$$

and there exists a point $r_0 \in (0, R]$ such that

$$\rho v(r_0) = u(r_0) \quad \text{and} \quad \rho v'(r_0) = u'(r_0). \tag{1.59}$$

Here, $r_0 = 0$ is impossible since $u'(0) = v'(0) = -M$ by (iii). By multiplying (1.49) for v with $\beta(\rho, v')\rho$ and subtracting this from (1.49) for u, we obtain

$$\begin{aligned} \tilde{W}''(u')\Delta u - \beta(\rho, v')\tilde{W}''(v')\Delta(\rho v) \quad &- \quad \frac{N-1}{r}\left[H(u')u' - H(\rho v')\rho v'\right] \\ &= \frac{G'(u)}{u} u - \beta(\rho, v')\frac{G'(v)}{v}\rho v. \end{aligned}$$

In a vicinity of r_0, this implies that

$$\begin{aligned} \tilde{W}''(u')\Delta(u - \rho v) - \frac{N-1}{r}&\left[H(u')u' - H(\rho v')\rho v'\right] \\ &\geq \frac{G'(u)}{u}(u - \rho v), \end{aligned} \tag{1.60}$$

provided that

$$-\beta(\rho, v')\frac{G'(v)}{v} \geq -\frac{G'(v)}{v} \geq -\frac{G'(u)}{u} \text{ and} \tag{1.61}$$

$$\beta(\rho, v')\tilde{W}''(v') \leq \tilde{W}''(u'), \tag{1.62}$$

for r near r_0, since $v \geq 0$ and $\Delta v \leq 0$. Inequality (1.61) is a consequence of (1.52), (1.54) the fact that $v < u$ near r_0 (because $0 < \rho v(r_0) = u(r_0)$ and $\rho > 1$). In order to prove (1.62), observe that since $u'(r_0) = \rho v'(r_0)$, we have that $\tilde{W}''(u'(r))$ is close to $\tilde{W}''(\rho v'(r))$ by continuity, as long as r is close enough to r_0. Thus (1.62) follows from (1.53), concluding the proof of (1.60).

Remembering that $u - \rho v \leq 0$ and $u(r_0) = \rho v(r_0)$ we infer that $u \equiv \rho v$ near r_0, using Hopf's maximum principle (note that by Taylor expansion, $H(u')u' - H(\rho v')\rho v' = c(r)(u' - \rho v')$ with a bounded coefficient function $c(r)$). By repeating the argument we get that $u \equiv \rho v$ everywhere on $(0, R)$. Since $u'(0) = v'(0) = -M$ by (iii), this implies that $\rho = 1$ and $u \equiv v$, a contradiction. □

Chapter 2

The singularly perturbed functional

In this chapter we study a singularly perturbed version of the functional E_0 in the radially symmetric case. It arises from E_0 in the following way:

$$E_\epsilon(u) := \int_{B_R(0)} \left[\frac{\epsilon}{2}(\Delta u)^2 + W(\nabla u) + G(u)\right] dx,$$

where $\epsilon > 0$ is a (small) parameter and $B_R(0)$ is the ball in $\mathbb{R}^N$ with radius $R > 0$ centered at 0. Here, the perturbed functional E_ϵ is defined on functions $u \in W^{2,2}(B_R(0)) \cap W_0^{1,p}(B_R(0))$.

The aim of this chapter is to obtain critical points for E_ϵ and their qualitative properties. We will also show that they converge to a nontrivial critical point of the limit functional E_0, as ϵ goes to zero. If W and G satisfy the assumptions of the uniqueness result in Chapter 1 (Theorem 1.6.1) then the singular limit is even characterized as the unique global minimizer of the limit energy E_0. Our method can be seen as a generalization of the one-dimensional case studied in [23, 25] to a radially symmetric problem in higher dimensions.

From a variational point of view, the main advantage of E_ϵ over E_0 is of course that it is convex in the derivatives of highest order and thus weakly lower semi-continuous, at least under appropriate growth ("subcritical", cf. Remark 2.1.3) and regularity assumptions on W and G. As to the question of critical points, the Euler–Lagrange equation of E_ϵ has the benefit of being an (strictly uniformly) elliptic equation, whereas the one corresponding to E_0 changes its local type from elliptic to hyperbolic, depending on the value of ∇u. The price we pay is that the equation for E_ϵ is of fourth order. In fact this is the main reason why we have to restrict ourselves to the case of radial symmetry, as our approach is based on a global bifurcation analysis which heavily relies on the maximum principle in order to separate global continua. As it is well known, elliptic equations of higher order or elliptic systems have to be of very special form in order to apply a maximum principle. However our equation does not have such a structure in general. In particular, we cannot rewrite it in the form of a second order system which is cooperatively coupled to use the maximum principle as in [20]. We overcome this problem by using information about the form of the solutions we focus on; properties which unfortunately are not available without the symmetry assumption, cf. Remark 2.3.13. Note also that it is far more difficult to obtain properties of a global minimizer of E_ϵ than it was in the case of E_0, because many of the rearrangements used in Chapter 1 for this purpose no longer work. This is due to the fact that in general the rearranged function has more energy than the original one or is not admissible at all, because of the ϵ-term in the functional which consists of second order derivatives. In particular the proof of radial symmetry of global minimizers fails. Here lies the main benefit of the method resting on global bifurcation, because the shape of critical points is essentially preserved along a continuum as a consequence of the maximum principle, which turns out to be essential to obtain compactness as $\epsilon \to 0$. The drawback is that we are unable to prove that the critical points on the bifurcating global continuum are global or at least local minimizers. An interesting byproduct of our discussion is the existence of a nontrivial, positive radially symmetric solution of the Euler–Lagrange equation even if the growth exponent p of W is critical or supercritical (i. e. $p = 2^*$ or $p > 2^*$, cf. Remark 2.1.3). In this case, the standard approach to the variational problem based on a straightforward application of the direct methods cannot be used to obtain a global minimizer, due to the lack of compactness of $W(\nabla u)$ (as a Nemytskii operator mapping $W^{2,2} \cap W^{1,p}$ into L^1).

For the bifurcation analysis we introduce a second parameter $\lambda \in \mathbb{R}$:

$$E_{\epsilon,\lambda}(u) := \int_{B_R(0)} \left[\frac{\epsilon}{2}(\Delta u)^2 + W(\lambda, \nabla u) + G(u)\right] dx, \qquad (E_{\epsilon,\lambda})$$

For us, the role of λ is of purely technical nature, which justifies choosing the dependance of λ in a specific way well suited to our purposes. The parameter value $\lambda = 0$ is interpreted as the one reproducing our "original" energy E_ϵ. Nevertheless, as the example of the stationary Cahn-Hilliard equation (cf. e. g. [22], and the references therein) shows, additional parameters like λ can naturally appear in physical models. In models for the elastic behavior of shape memory materials the temperature actually plays the role of λ: The energy density $W(\lambda, Du)$ (in this case depending on the deformation matrix $Du \in \mathbb{R}^{3\times 3}$) has two or more wells at low temperature λ, whereas sufficient heating deforms W back to the classical one-well shape of standard nonlinear elasticity, cf. [3].

2.1 Preliminaries

We study solutions $u \in W^{2,2}(B_R(0)) \cap W_0^{1,p}(B_R(0))$ of the (weak) Euler–Lagrange equation corresponding to $E_{\epsilon,\lambda}$: For every test function $\varphi \in W^{2,2}(B_R(0)) \cap W_0^{1,p}(B_R(0))$,

$$\int_{B_R(0)} \epsilon\Delta u\Delta\varphi + \nabla W(\lambda,\nabla u)\cdot\nabla\varphi + G'(u)\varphi\,dx = 0. \qquad (EL_\epsilon)$$

Our assumptions on W resemble those used in the first chapter, apart from the newly introduced parameter λ, additional regularity assumptions and the shape condition (W_3). Generally speaking, the assumptions below always imply the corresponding ones in the first chapter, whereas the converse is not always true. There is one exception to the rule, however, namely (G_1), but on the other hand we have to replace this one by stronger conditions in context of the singular limit process. For every $\xi \in \mathbb{R}^N$, we assume the following:

(Dependance on λ)	$W(\lambda,\xi) = \frac{1}{2}\lambda^2\lvert\xi\rvert^2 + W_0(\xi)$,	(W_λ)
(Regularity)	$W_0 : \mathbb{R}^N \to \mathbb{R}$ is of class C^2,	(W_0)
(Coercivity)	$\nabla W_0(\xi)\cdot\xi \geq \nu_1\lvert\xi\rvert^p - C$,	(W_1)
(Growth)	$\lvert\nabla W_0(\xi)\rvert \leq \nu_2\lvert\xi\rvert^{p-1} + C$,	(W_2)
(Symmetry)	$W_0(\xi) = \tilde{W}_0(\lvert\xi\rvert)$, where $\tilde{W}_0 : \mathbb{R} \to \mathbb{R}$ is an even function of class C^2. Notation: $\tilde{W}(\lambda,t) := \frac{\lambda^2}{2}t^2 + \tilde{W}_0(t),\ t \in \mathbb{R}$.	(W_{sym})
(Shape)	$\frac{1}{t}\tilde{W}_0'(t) \leq \tilde{W}_0''(t)$ for every $t \in \mathbb{R}\setminus\{0\}$.	(W_3)

Here, $p > 1$, $\nu_2 \geq \nu_1 > 0$, and C are real constants. The assumptions on G are as follows:

(*Regularity*)	$G : \mathbb{R} \to \mathbb{R}$ is of class C^2,	(G_0)
(*Shape*)	$G' \leq 0$ on $[0,\infty)$ and $G'(0) = 0$	(G_1)
(*Growth*)	$\lvert G'(\mu)\rvert \leq \nu_3\lvert\mu\rvert^{\tilde{p}-1} + C$ for every $\mu \in \mathbb{R}$,	(G_2)

where $\tilde{p} \in [1,p)$ and $\nu_3 \geq 0$ are constants. If both u and the test function φ are radially symmetric, (EL_ϵ) can be rewritten as

$$\int_0^R \left[\epsilon\Delta u\Delta\varphi + \tilde{W}'(\lambda,u')\varphi' + G'(u)\varphi\right] r^{N-1}\,dr = 0, \qquad (EL_{r,\epsilon})$$

with $\Delta u = u'' + (N-1)r^{-1}u'$, where $\tilde{W}'$ denotes the derivative of $\tilde{W}$ with respect to its second variable: $\tilde{W}'(\lambda, t) = \lambda^2 t + \tilde{W}_0'(t)$ according to (W_λ) and (W_{sym}). Moreover, by virtue of the regularity of radially symmetric solutions asserted in Proposition 2.1.2 below, u satisfies (EL_ϵ) even in the strong sense:

$$\epsilon \Delta^2 u - \frac{1}{r^{N-1}} \frac{d}{dr} \left[r^{N-1} \tilde{W}'(\lambda, u') \right] + G'(u) = 0 \text{ on } (0, R). \qquad (EL^s_{r,\epsilon})$$

Note also that this equation is the radial version of the strong form of (EL_ϵ) since

$$\frac{1}{r^{N-1}} \frac{d}{dr} \left[r^{N-1} \tilde{W}'(\lambda, u'(r)) \right] = \operatorname{div} \left[\nabla W(\lambda, \nabla u(x)) \right] \text{ whenever } |x| = r,$$

due to (W_{sym}).

Remark 2.1.1. (i) The main point in our choice of the behavior in λ is that $W(\lambda, \cdot)$ becomes strictly convex for large λ, at least if W_0 is locally convex outside a bounded set. Note also that $W(0, \xi) = \tilde{W}_0(|\xi|)$, i. e. the parameter value $\lambda = 0$ corresponds to the "original" function W.

(ii) The additional regularity of W_0 and G (as opposed to the first chapter) enables us to differentiate (EL_ϵ) with respect to u.

(iii) Assumption (W_{sym}) implies that $\nabla W_0(0) = 0$. Since we also have that $G'(0) = 0$ by (G_1), the Euler–Lagrange equation (EL_ϵ) always admits the "trivial" solution $u \equiv 0$, for every $\epsilon > 0$ and $\lambda \in \mathbb{R}$.

(iv) Assumption (W_3) noticeably restricts the class of admissible functions W, in particular, $\tilde{W}_0$ can have at most two wells. On the other hand, any function $\tilde{W}_0$ such that $\tilde{W}_0'$ is convex on $\mathbb{R}^+$ and concave on $\mathbb{R}^-$ satisfies (W_3). Our assumptions in the first chapter admit a much larger class of functions $\tilde{W}_0$. We will use (W_3) only once, at an essential point in the proof of Proposition 2.3.12, where the maximum principle is used to show that the sign of solutions is preserved along connected sets of nontrivial solutions.

Observe that solutions of (EL_ϵ) have additional regularity:

Proposition 2.1.2. *Assume (W_λ), (W_0)–(W_2), (W_{sym}), and (G_0)–(G_2). Then any solution $u \in W^{2,2}(B_R(0)) \cap W_0^{1,p}(B_R(0))$ of (EL_ϵ) is an element of $W^{3,P/(P-1)}(B_R(0))$, $P := \max\{p, 2\}$, and satisfies the corresponding a-priori estimate*

$$\|u\|_{W^{3,P/(P-1)}(B_R(0))} \leq C \frac{\lambda^2 + 1}{\epsilon} \left(\|u\|^{P-1}_{W^{1,P}(B_R(0))} + 1 \right), \qquad (2.1)$$

with a constant C independent of u, ϵ and λ. Furthermore, any radially symmetric solution is of class C^4 on $\overline{B_R(0)} \setminus \{0\}$ and satisfies the natural boundary condition

$$\Delta u(R) = \left(\frac{\partial^2}{\partial r^2} + \frac{N-1}{r} \frac{\partial}{\partial r} \right) u(r) \Big|_{r=R} = 0. \qquad (2.2)$$

Proof. The weak differentiability of u up to third order and the corresponding a–priori estimate are an immediate consequence of the elliptic regularity theory in Sobolev spaces, cf. e. g. [1]; note that

$$\|\nabla\dot{W}(\lambda,\nabla u)\|_{L^{P/(P-1)}} \leq C(\lambda^2+1)\left(\|u\|^{P-1}_{W^{1,P}(B_R(0))}+1\right)$$

with a constant C independent of u, ϵ and λ, by virtue of (W_λ) and (W_2).

As to the radial symmetric case, first observe that due to the one dimensional Sobolev imbedding u is of class C^2 on $(0,R]$. Since W is of class C^2, the asserted C^4-regularity now follows from the Fundamental Lemma of Du Bois–Reymond. Thus u is a classical solution on $B_R(0)\setminus\{0\}$, and the natural boundary condition follows from (EL_ϵ) with a radially symmetric test function $\varphi \in C^\infty(0,R]$ satisfying $0 \notin \operatorname{supp}\varphi$, $\varphi(R)=0$ and $\partial_r\varphi(R)=1$. □

Remark 2.1.3. All solutions also have to be elements of $W^{3,2^*/(P-1)}(B_R(0))$, where $P := \max\{p,2\}$ as before and $2^* := 2N/(N-2)$ is the critical Sobolev exponent. Furthermore,

$$\|u\|_{W^{3,2^*/(P-1)}(B_R(0))} \leq \frac{C}{\epsilon}\left(\|u\|^{P-1}_{W^{1,2^*}(B_R(0))}+1\right).$$

Moreover if p is subcritical in the sense that $p<2^*$, this inequality can be used as a starting point of a bootstrap argument based on elliptic a–priori estimates, which yields that u is bounded in $W^{3,q}$ for arbitrary $q>1$, the bound depending on ϵ and λ. However we will not follow this path since in our setting appropriate estimates for radial solutions are available even for supercritical p as expounded in Section 2.2.

The regularity of radial solutions can also be used to obtain a natural "boundary" condition at 0 and an integrated version of $(EL^s_{r,\epsilon})$:

Proposition 2.1.4. *Assume (W_λ), (W_0)–(W_2), (W_{sym}), and (G_0)–(G_2). Then any radially symmetric solution $u \in W^{2,2}(B_R(0)) \cap W^{1,p}_0(B_R(0))$ of (EL_ϵ) satisfies the "boundary" condition*

$$\epsilon r^{N-1}\frac{d}{dr}\Delta u(r) - r^{N-1}\tilde{W}'(\lambda,u') \to 0 \quad \text{as } r\to 0. \tag{2.3}$$

Furthermore, u solves

$$\epsilon r^{N-1}\frac{d}{dr}\Delta u - r^{N-1}\tilde{W}'(\lambda,u') + \int_0^r G'(u(s))s^{N-1}\,ds = 0 \tag{2.4}$$

for every $r \in (0,R]$.

Proof. Choose a radius $r_0 \in (0, R)$ and a radially symmetric test function $\varphi \in C_0^\infty(B_R(0))$ such that $\varphi(r) = 1$ for every $r \leq r_0$. Then for every $s \in (0, r_0)$ $(EL_{r,\epsilon})$ yields

$$0 = \int_s^R \Big[\epsilon \Delta u \Delta \varphi + \tilde{W}'(\lambda, u')\varphi' + G'(u)\varphi\Big] r^{N-1}\,dr + \int_0^s G'(u) r^{N-1}\,dr$$

and integrating by parts we conclude that

$$0 = \epsilon r^{N-1} \frac{d}{dr}\Delta u(r) - r^{N-1}\tilde{W}'(\lambda, u')\Big|_{r=s} + \int_0^s G'(u) r^{N-1}\,dr,$$

since u is a strong solution of $(EL_{r,\epsilon})$ in (s, R) as a consequence of its regularity observed in Proposition 2.1.2. Furthermore, $G'(u)r^{N-1}$ is integrable on $(0, R)$ and thus

$$\int_0^s G'(u) r^{N-1}\,dr \to 0 \quad \text{as } s \to 0,$$

which implies (2.3). To prove (2.4), integrate $(EL_{r,\epsilon}^s)$ from s to x with $0 < s < x \leq R$ which yields

$$\begin{aligned} 0 &= \int_s^x \epsilon \frac{d}{dr}\left[r^{N-1}\frac{d}{dr}\Delta u - r^{N-1}\tilde{W}'(\lambda, u')\right] + r^{N-1}G'(u)\,dr \\ &= \left[\epsilon r^{N-1}\frac{d}{dr}\Delta u - r^{N-1}\tilde{W}'(\lambda, u')\right]_{r=s}^{r=x} + \int_s^x G'(u) r^{N-1}\,dr. \end{aligned}$$

Now pass to the limit as $s \to 0$ and use (2.3) to obtain (2.4). □

2.2 A priori estimates

The estimates derived in this section serve two purposes: On the one hand, we need estimates to exclude the possibility of a blow up of global solution branches, on the other hand estimates independent of ϵ are a vital tool for the singular limit process, i. e. to obtain a limit solution as $\epsilon \to 0$. We start with a very simple estimate which mimics the coercivity of the energy and is available even in situations without symmetry:

Proposition 2.2.1. *Assume that W and G are of class C^1 and satisfy (W_λ), (W_1), (W_2) and (G_2). Then there is a constant $C > 0$ such that any solution $u \in W^{2,2}(B_R(0)) \cap W_0^{1,p}(B_R(0))$ of (EL_ϵ) satisfies*

$$\epsilon \int_{B_R(0)} (\Delta u)^2\,dx + \lambda^2 \int_{B_R(0)} |\nabla u|^2\,dx + \int_{B_R(0)} |\nabla u|^p\,dx \leq C.$$

Here, the constant C does not depend on u, ϵ and λ. In particular,

$$\|u\|_{W^{2,2}} \leq \frac{C'}{\epsilon} \qquad \textit{and} \qquad \|u\|_{W^{1,p}} \leq C',$$

with another constant $C' > C$.

Proof. Using the solution u as a test function in (EL_ϵ) we obtain

$$\epsilon \int (\Delta u)^2 \, dx + \int \left[\lambda^2 |\nabla u|^2 + \nabla W_0(\nabla u) \cdot \nabla u\right] \, dx = -\int G'(u) u \, dx.$$

Here, the domain of integration always is $B_R(0)$. As a consequence of (W_1) and (G_2) we infer that

$$\epsilon \int (\Delta u)^2 \, dx + \int \lambda^2 |\nabla u|^2 \, dx + \nu_1 \int |\nabla u|^p \, dx \leq \nu_3 \int |u|^{\tilde{p}} \, dx + C_1,$$

with a constant C_1 independent of u, ϵ and λ. Using both Poincaré's and Hölder's inequality we get

$$\epsilon \int (\Delta u)^2 \, dx + \int \lambda^2 |\nabla u|^2 \, dx + \nu_1 \int |\nabla u|^p \, dx \leq C_2 \left(\int |\nabla u|^p \, dx \right)^{\tilde{p}/p} + C_1.$$

Since the exponent $\tilde{p}/p$ is less than one, we can find a constant $\tilde{C}$ such that

$$-\frac{\nu_1}{2} t + C_2 t^{\tilde{p}/p} + C_1 \leq \tilde{C} \quad \text{for every } t \geq 0,$$

and thus, with $t := \int |\nabla u|^p \, dx$,

$$\epsilon \int (\Delta u)^2 \, dx + \int \lambda^2 |\nabla u|^2 \, dx + \frac{\nu_1}{2} \int |\nabla u|^p \, dx \leq \tilde{C},$$

which implies the assertion. □

In a very similar way one can also show that the only solution of (EL_ϵ) is the trivial one, provided that either λ or ϵ is large:

Proposition 2.2.2. *Assume (W_λ), (W_0)–(W_2), (G_0) and (G_2) as well as $\nabla W_0(0) = 0$ and $G'(0) = 0$. Then there is a constant $\zeta > 0$ such that $u \equiv 0$ is the unique solution $u \in W^{2,2}(B_R(0)) \cap W_0^{1,p}(B_R(0))$ of (EL_ϵ) whenever $\lambda^2 \geq \zeta$ or $\epsilon \geq \zeta$.*

Proof. Since W_0 is of class C^2 and $\nabla W_0(0) = 0$, (W_1) implies

$$\nabla W_0(\xi) \cdot \xi \geq \nu_1 |\xi|^p - \tilde{C} |\xi|^2, \text{ for every } \xi \in \mathbb{R}^N,$$

with a new constant $\tilde{C} \in \mathbb{R}$. Analogously, (G_2) can be replaced by

$$|G'(\mu)\mu| \leq \nu_3 |\mu|^{\tilde{p}} + \tilde{C}\mu^2 \text{ for every } \mu \in \mathbb{R}.$$

By testing (EL_ϵ) with the solution u and employing the estimates above, we arrive at

$$\begin{aligned} \epsilon \int (\Delta u)^2\, dx + \lambda^2 \int |\nabla u|^2\, dx + \nu_1 \int |\nabla u|^p\, dx \\ \leq \tilde{C} \int \left(|\nabla u|^2 + u^2\right) dx + \nu_3 \int |u|^{\tilde{p}}\, dx. \end{aligned} \tag{2.5}$$

By virtue of Poincaré's inequality there exists a constant $\eta > 0$ such that

$$\int (\Delta u)^2\, dx \geq \eta \int |\nabla u|^2\, dx.$$

Thus (2.5) implies

$$(\lambda^2 + \epsilon\eta) \int |\nabla u|^2\, dx + \nu_1 \int |\nabla u|^p\, dx \leq \tilde{C} \int \left(|\nabla u|^2 + u^2\right) dx + \nu_3 \int |u|^{\tilde{p}}\, dx.$$

Again by Poincaré's inequality, this time used to estimate the $L^{\tilde{p}}$-, respectively, L^2-norm of u on the right hand side in terms of the corresponding norm of the gradient, this inequality can be further reduced to

$$(\lambda^2 + \epsilon\eta) \int |\nabla u|^2\, dx + \nu_1 \int |\nabla u|^p\, dx \leq \tilde{C}_1 \int |\nabla u|^2\, dx + \tilde{C}_2 \int |\nabla u|^{\tilde{p}}\, dx. \tag{2.6}$$

Moreover, since $p > \tilde{p}$, we have the elementary estimate

$$\zeta t^2 + \frac{\nu_1}{2} |t|^p \geq \tilde{C}_1 t^2 + \tilde{C}_2 |t|^{\tilde{p}} \text{ for every } t \in \mathbb{R},$$

provided the constant $\zeta > 0$ has been chosen large enough. Consequently, (2.6) implies

$$\frac{\nu_1}{2} \int |\nabla u|^p\, dx \leq 0 \quad \text{if } \lambda^2 + \epsilon\eta \geq \zeta,$$

so that $u \equiv 0$ as claimed. □

The benefits of our next result are twofold: On the one hand, it allows us to derive a suitable a–priori estimate for bifurcation analysis of the next section without having to assume that p is subcritical. On the other hand, it is used as a tool for the singular limit process. In particular, note that it is uniform in ϵ, which is slightly surprising because any approach based on elliptic regularity theory and the corresponding linear a-priori estimates inevitably produces a dependence on ϵ since in this context the elliptic leading term containing the derivatives of highest order would be used to dominate other terms of lower order. Quite contrary to this concept, in the proof below the ϵ-term is treated as a perturbation which has to be estimated. However we still make use of its ellipticity.

Proposition 2.2.3. *Assume (W_λ), (W_0)–(W_2), (W_{sym}), (G_0) and (G_2). Furthermore let $u \in W^{2,2}(B_R(0)) \cap W_0^{1,p}(B_R(0)) \cap W^{1,\infty}(B_R(0))$ be a radially symmetric solution of (EL_ϵ). Then there is a constant $C \geq 0$ independent of r, u, ϵ and λ such that*

$$|u'(r)| \leq C \quad \text{for every } r \in (0, R).$$

Proof. Our proof is based on a comparison argument applied to the integrated version (2.4) of the Euler–Lagrange equation which we derived in Proposition 2.1.4. The function u' is compared with a family of maps

$$h_{\alpha,\beta}(r) := \alpha r^{-\beta} + |u'(R)|,$$

where $\alpha \geq 0$ and $\beta > 0$ are real parameters. We choose $\alpha_0 = \alpha_0(\beta)$ in such a way that $h_{\alpha_0,\beta}$ touches $|u'|$ from above:

$$\alpha_0(\beta) := \min\{\alpha \geq 0 \mid |u'| \leq h_{\alpha,\beta} \text{ on } (0, R)\}.$$

By construction,

$$h_\beta := h_{\alpha_0(\beta),\beta} \geq |u'| \quad \text{on } (0, R).$$

Furthermore there exists a point $r_0 = r_0(\beta) \in (0, R]$ such that

$$|u'(r_0)| = h_\beta(r_0) > 0.$$

Now consider the case $u'(r_0) < 0$ (the reasoning for $u'(r_0) > 0$ is analogous). First we show that $r_0 = R$ is impossible for $\beta < N - 1$: An elementary computation yields

$$h'_\beta(R) + \frac{N-1}{R} h_\beta(R) = \alpha_0(N - 1 - \beta)R^{-\beta-1} + \frac{N-1}{R}|u'(R)| > 0,$$

at least if either $u'(R) \neq 0$ or $\alpha_0 \neq 0$ (otherwise $|u'| \leq h_{0,\beta} \equiv 0$ anyway). Moreover, $\Delta u(R) = 0$ by Proposition 2.1.2, so that

$$h'_\beta(R) + \frac{N-1}{R} h_\beta(R) + \left(u''(R) + \frac{N-1}{R} u'(R)\right) > 0.$$

If $r_0 = R$ then $h_\beta(R) = -u'(R)$ by definition of r_0, and we infer that

$$h'_\beta(R) + u''(R) > 0,$$

which contradicts the fact that $h_\beta \geq -u'$ on $(0, R)$. Therefore we know that $r_0 \in (0, R)$ and we additionally get

$$u''(r_0) = -h'_\beta(r_0) \quad \text{and} \quad u'''(r_0) \geq -h''_\beta(r_0),$$

whereby

$$\begin{aligned}
\frac{d}{dr}\Delta u\Big|_{r=r_0} &\geq -h''_\beta(r_0) - \frac{N-1}{r_0}h'_\beta(r_0) + \frac{N-1}{r_0^2}h_\beta(r_0) \\
&= \alpha_0\left[(N-1-\beta)(\beta+1)\right] r_0^{-\beta-2} + \frac{N-1}{r_0^2}\left|u'(R)\right| \\
&\geq 0 \quad \text{provided that } \beta < N-1.
\end{aligned}$$

Thus (2.4) implies that

$$-r_0^{N-1}\tilde{W}'(\lambda, u'(r_0)) \leq \int_0^{r_0} \left|G'(u(s))\right| s^{N-1}\, ds \quad (\text{if } u'(r_0) < 0). \tag{2.7}$$

The left hand side of (2.7) can be estimated using the coercivity condition (W_1):

$$-\tilde{W}'(\lambda, t) \geq -\tilde{W}'_0(t) \geq c\,|t|^{p-1} - C \quad \text{for every } t \leq 0,$$

with constants $c > 0$, $C \geq 0$. Since $-u'(r_0) = h_\beta(r_0) = \alpha_0 r_0^{-\beta} + |u'(R)|$, (2.7) yields

$$r_0^{N-1}\left[c\left(\alpha_0 r_0^{-\beta} + |u'(R)|\right)^{p-1} - C\right] \leq \int_0^{r_0} \left|G'(u(s))\right| s^{N-1}\, ds. \tag{2.8}$$

Doing the analogous computation for the case $u'(r_0) > 0$, we also end up with (2.8).

We continue with an estimate for the right hand side of (2.8). From now on, we assume that $\beta < 1/2$. Then

$$\begin{aligned}
|u(s)| &\leq \int_0^R |u'(t)|\, dt \leq \int_0^R |h_\beta(t)|\, dt \\
&= \frac{1}{1-\beta}\left[\alpha_0 R^{1-\beta} + R\,|u'(R)|\right] \leq 2\left[\alpha_0(1+R) + R\,|u'(R)|\right]
\end{aligned}$$

for every $s \in (0, R)$. Thus, by virtue of (G_2),

$$|G'(u(s))| \leq C_1(\alpha_0 + |u'(R)|)^{\tilde{p}-1} + C_2,$$

with constants $C_1, C_2 \geq 0$ which depend on R but not on u, β, α_0 or $u'(R)$. Consequently, (2.8) reduces to

$$c\left(\alpha_0 r_0^{-\beta} + |u'(R)|\right)^{p-1} - C \leq \frac{r_0}{N}\left[C_1(\alpha_0 + |u'(R)|)^{\tilde{p}-1} + C_2\right].$$

Using that $r_0 \leq R$ and $r_0^{-\beta} \geq R^{-\beta} \geq \min\{1, R^{-1/2}\}$, we can find new constants $C_3, C_4 \geq 0$ such that

$$c\,(\alpha_0 + |u'(R)|)^{p-1} \leq C_3(\alpha_0 + |u'(R)|)^{\tilde{p}-1} + C_4.$$

Since $c > 0$ and $p > \tilde{p}$ this implies that both $\alpha_0 = \alpha_0(\beta)$ and $|u'(R)|$ are bounded for $\beta \in (0, 1/2)$, uniformly in u, β, ϵ and λ. Thus

$$|u'(r)| \leq \inf_{\beta \in (0,1/2)} h_\beta(r) \leq \sup_{\beta \in (0,1/2)} \alpha_0(\beta) + |u'(R)|$$

is uniformly bounded as well. □

An immediate consequence of the bound in $W^{1,\infty}$ is

Corollary 2.2.4. *Assume* (W_λ), (W_0)*–*(W_2), (W_{sym}), (G_0) *and* (G_2)*. Furthermore fix* $\alpha \in (0,1)$ *and let* $u \in W^{1,\infty}(B_R(0)) \cap W^{2,2}(B_R(0))$, $u = 0$ *on* $\partial B_R(0)$*, be a radially symmetric solution of* (EL_ϵ)*. Then* $u \in W^{4,q}(B_R(0))$ *for every* $q \in (1, \infty)$ *and*

$$\|u\|_{W^{4,q}} \leq C.$$

with a constant $C = C(\epsilon, q) \geq 0$ *which is independent of* r*,* u *and* λ *and bounded for bounded* ϵ^{-1}*. In particular,* $u \in C^{3,\alpha}(B_R(0))$ *and*

$$\|u\|_{C^{3,\alpha}} \leq \tilde{C}.$$

with another constant $\tilde{C} = \tilde{C}(\epsilon, \alpha) \geq 0$ *which is independent of* r*,* u *and* λ *and bounded for bounded* ϵ^{-1}*. Moreover,* u *is a (strong) solution of* $(EL^s_{r,\epsilon})$ *on* $B_R(0)$ *(the origin included!).*

Proof. By Proposition 2.2.3, all radially symmetric solutions with bounded first derivatives are uniformly bounded in $W^{1,\infty}$. Therefore one can use elliptic regularity theory in Sobolev spaces (cf. e. g. [1]) to infer that those solutions belong to $W^{3,q}$ and are uniformly bounded therein (the bound of course now depending on ϵ and q), for arbitrary $q \in (1, \infty)$. Since $W^{3,q}(B_R(0))$ is continuously imbedded into $C^{2,\alpha}(B_R(0))$ if q is large enough, we can consider (EL_ϵ) as an equation for Δu in following sense: For every test function $\varphi \in C_0^\infty(B_R(0))$,

$$\int_{B_R(0)} \epsilon \Delta u \Delta \varphi \, dx = \int_{B_R(0)} \left[\operatorname{div} \nabla W(\lambda, \nabla u) - G'(u)\right] \varphi \, dx,$$

with a right hand side div $\nabla W(\lambda, \nabla u) - G'(u)$ which is bounded in L^∞ (remember that W is of class C^2). Thus another application of regularity theory yields that $u \in W^{4,q}(B_R(0))$ for every $q > 1$, and u is bounded in the corresponding norm as claimed. □

2.3 Bifurcation analysis

The key tool in the analysis of this section is the topological degree. For the application of the Leray–Schauder degree we have to reformulate (EL_ϵ) to an equation of the form

$$F_\epsilon(\lambda, u) = 0 \tag{2.9}$$

in such a way that $F_\epsilon(\lambda, \cdot)$ is a compact perturbation of the identity in a suitable Banach space X. For this purpose fix $\alpha \in (0,1)$ (arbitrarily), choose a number $P > 1$ large enough so that $W^{4,P}(B_R(0))$ is compactly imbedded into $C^{3,\alpha}(B_R(0))$ and let

$$X := \left\{ u \in C^{3,\alpha}(B_R(0)) \;\middle|\; \begin{array}{l} u \text{ is radially symmetric,} \\ u = \Delta u = 0 \text{ on } \partial B_R(0) \end{array} \right\}, \tag{2.10}$$

endowed with the norm of $C^{3,\alpha}(B_R(0))$. The function F_ϵ is defined by

$$\begin{aligned} &F_\epsilon : \mathbb{R} \times X \to X, \\ &F_\epsilon(\lambda, u) := u + \frac{1}{\epsilon} K(\lambda, u), \quad \text{where} \\ &K(\lambda, u) := \Delta^{-2} \left(-\operatorname{div} \left[\nabla W(\lambda, \nabla u) \right] + G'(u) \right). \end{aligned} \tag{2.11}$$

Here, the squared inverse Laplacian Δ^{-2} is interpreted in the following way: For $a \in L^P(B_R(0))$ we define the function $\Delta^{-2} a$ as the unique solution $v_2 \in W^{4,P}(B_R(0)) \cap W_0^{1,P}(B_R(0))$ of

$$\Delta v_2 = v_1 \quad \text{in } B_R(0),$$

where the function $v_1 \in W^{2,P}(B_R(0)) \cap W_0^{1,P}(B_R(0))$ is the unique solution of

$$\Delta v_1 = a \quad \text{in } B_R(0).$$

By the compact imbedding of $W^{4,P}$ into $C^{3,\alpha}$, by the fact that radial symmetry is preserved (in particular due to (W_{sym})) and by the continuity of W in λ, we conclude that

$$K : \mathbb{R} \times X \to X \quad \text{is well defined and completely continuous.}$$

Remark 2.3.1. The number P plays a purely auxiliary role and the definition of K does not depend on P. In particular, K maps X to $W^{4,P}$ for arbitrarily large P.

We observe the following connection of (2.9) and (EL_ϵ):

Proposition 2.3.2. *Assume* (W_λ), (W_0), (W_2), (W_{sym}), (G_0) *and* (G_2). *Then any solution of (2.9) also solves* (EL_ϵ). *Vice versa, any radially symmetric solution of* (EL_ϵ) *belonging to* $W^{1,\infty}(B_R(0))$ *(which is always the case if* p *is subcritical, i.e.* $p < 2^*$, *c.f. remark 2.1.3) is in* X *and solves (2.9).*

Proof. The claimed regularity of solutions of (EL_ϵ) of class $W^{1,\infty}$ is due to Corollary 2.2.4. The remaining assertions directly follow from the definition of F_ϵ. □

Next we collect a few basic regularity properties of F_ϵ:

Proposition 2.3.3. *Assume* (W_λ), (W_0), (W_2), (W_{sym}), (G_0) *and* (G_2). *If* W_0 *and* G *are of class* C^{k+1} *for a* $k \in \mathbb{N}$ *then the function* F *is of class* C^k *for every* $(\epsilon, \lambda, u) \in (0, \infty) \times \mathbb{R} \times X$.

Proof. The proof is straightforward, the details are left to the reader. □

Traditionally, bifurcation is studied for problems depending on a single parameter. However in our case, two parameters are involved, namely, λ and ϵ. Therefore two different points of view are possible, namely keeping ϵ fixed and varying λ or vice versa. In the following, each assertion is split into two parts accordingly.

2.3.1 Local bifurcation from the trivial solution

For every $\epsilon > 0$ and $\lambda \in \mathbb{R}$, the equation $F_\epsilon(\lambda, u) = 0$ admits the trivial solution $u \equiv 0$, since $\nabla W(\lambda, 0) = 0$ and $G'(0) = 0$ by (W_1), (W_{sym}) and (G_1). The points admitting bifurcation from the trivial solution are given by those values of λ and ϵ where the linearization of $F_\epsilon(\lambda, \cdot)$ at zero becomes singular. Thus we are led to seek nontrivial solutions $v \in X$ of

$$D_u F_\epsilon(\lambda, 0)v = 0, \tag{2.12}$$

where

$$D_u F_\epsilon(\lambda, 0)v = v + \frac{1}{\epsilon}\Delta^{-2}\left(-\tilde{W}''(\lambda, 0)\,\mathrm{div}\,\nabla v + G''(0)v\right), \tag{2.13}$$

since by (W_{sym}), $D\nabla W(\lambda,0) = \tilde{W}''(\lambda,0)Id$. Before we continue to derive results specific to our setting, we recall a general asymptotic property of nontrivial solutions bifurcating from the trivial line:

Lemma 2.3.4. *Let X be a Banach space. Assume that the function $F : \mathbb{R}^+ \times \mathbb{R} \times X \to X$, $(\epsilon,\lambda,u) \mapsto F_\epsilon(\lambda,u)$ is Fréchet-differentiable with respect to u, and its derivative is continuous in (ϵ,λ,u). Furthermore, suppose that $F_\epsilon(\lambda,0) = 0$ for every $\lambda \in \mathbb{R}$ and $\epsilon > 0$ and let $(\varepsilon,\Lambda) \in \mathbb{R}^+ \times \mathbb{R}$ be a point where $K := \ker D_uF_\varepsilon(\Lambda,0) \neq \{0\}$ and $D_uF_\varepsilon(\Lambda,0) : X \to X$ is a Fredholm operator of index zero. Moreover we assume that there is a closed complement subspace H of K so that $X = K \oplus H$ (in the case of X defined as in (2.10), for example choose the orthogonal complement with respect to the L^2 scalar product). If $(\epsilon_n,\lambda_n,u_n)$ is a sequence of nontrivial solutions ($u_n \neq 0$) converging to $(\varepsilon,\Lambda,0)$ then the normalized sequence $u_n/\left\|u_n\right\|_X$ is relatively compact in X and the limit of any convergent subsequence is a unit vector in K.*

Proof. Taylor expansion yields

$$0 = F_{\epsilon_n}(\lambda_n,u_n) = D_uF_{\epsilon_n}(\lambda_n,0)u_n + o(\|u_n\|),$$

where $o(\|u_n\|)/\|u_n\| \to 0$ as $\|u_n\| \to 0$, uniformly in ϵ_n and λ_n. Thus

$$D_uF_{\epsilon_n}(\lambda_n,0)\frac{u_n}{\|u_n\|} \to 0 \text{ as } n \to \infty.$$

By continuity of D_uF in ϵ and λ we further infer that

$$D_uF_\varepsilon(\Lambda,0)\frac{u_n}{\|u_n\|} \to 0 \text{ as } n \to \infty;$$

note that $u_n/\|u_n\|$ is bounded. According to the decomposition of X, u_n can be written as

$$u_n = v_n + w_n, \text{ with corresponding sequences } v_n \in K \text{ and } w_n \in H,$$

and thus

$$D_uF_\varepsilon(\Lambda,0)\frac{w_n}{\|u_n\|} \to 0 \text{ as } n \to \infty.$$

Since w_n belongs to H and $D_uF_\varepsilon(\Lambda,0)$ has a bounded inverse if restricted to H (remember that $D_uF_\varepsilon(\Lambda,0)$ is a Fredholm operator of index 0, and H is a closed complement of the kernel K), this implies that

$$\frac{w_n}{\|u_n\|} \to 0 \text{ as } n \to \infty.$$

In particular, if a subsequence of $u_n/\|u_n\|$ converges then its limit has to belong to K. It remains to show that $u_n/\|u_n\|$, or, equivalently, $v_n/\|u_n\|$, is relatively compact. This is clear because K is finite dimensional and $v_n/\|u_n\|$ is bounded since the projection of X onto K along H is continuous. □

With the help of representation (2.13), it is not difficult to see that every solution $v \in X$ of (2.12) is a strong solution of

$$\begin{aligned} &\epsilon\Delta^2 v - \tilde{W}''(\lambda,0)\Delta v + G''(0)v = 0 \quad \text{in } B_R(0), \\ &v = 0 \text{ and } \Delta v = 0 \text{ on } \partial B_R(0). \end{aligned} \tag{2.14}$$

Here, the Dirichlet boundary condition on Δv is a consequence of the way we defined the action of Δ^{-2} in the definition of F_ϵ. It is remarkable that all the derivatives of v occurring in (2.14) are expressed in terms of the Laplacian. Because of this property we use a Fourier expansion with respect to the radially symmetric eigenfunctions of the Laplacian to determine the solutions of (2.14).

Proposition 2.3.5. *Assume (W_λ), (W_0)–(W_2), (W_{sym}) and (G_0)–(G_2). Moreover, let $0 < \mu_0 < \mu_1 < \ldots$ denote the eigenvalues of the radially symmetric negative Laplacian on $B_R(0)$ subject to Dirichlet boundary conditions and v_0, v_1, ... the corresponding radially symmetric eigenfunctions, normalized in $L^2(B_R(0))$. Then (2.14) (or equivalently, (2.12)) has a nontrivial solution if and only if there is a $k \in \mathbb{N}_0$ such that the characteristic equation*

$$\epsilon\mu_k^2 + \tilde{W}''(\lambda,0)\mu_k + G''(0) = 0 \tag{2.15}$$

is satisfied. The kernel of $D_uF_\epsilon(\lambda,0)$, i. e. the solution space of (2.14), is spanned by the set

$$L = L(\epsilon,\lambda) := \{v_k \mid \mu_k \text{ solves (2.15)}\}\,.$$

Its dimension, the number of elements of L, is always less or equal to one. Furthermore we have the following:

(i) (Useful for bifurcation analysis with ϵ fixed) The set

$$B_\epsilon := \{\lambda \in \mathbb{R} \mid \dim\ker D_uF_\epsilon(\lambda,0) > 0\}$$

is finite and has the following structure:

$$B_\epsilon = \{\pm\lambda_k \mid 0 \le k \le M-1\},$$

where $M = M(\epsilon) \in \mathbb{N}_0$ is the number of the $k \in \mathbb{N}_0$ such that (2.15) is solvable and $\lambda_k = \lambda_k(\epsilon) \ge 0$ is the unique nonnegative solution of (2.15) corresponding to μ_k and ϵ. For fixed ϵ, λ_k is strictly decreasing in k, and for fixed k, $\lambda_k(\epsilon)$ is strictly decreasing in ϵ. The number of elements of B_ϵ (or $M(\epsilon)$, respectively) increases as ϵ decreases. Furthermore $M(\epsilon) \to \infty$ as $\epsilon \to 0$ provided that either $\tilde{W}_0''(0) = 0$ and $G''(0) < 0$ or $\tilde{W}_0''(0) < 0$ (then, in particular, B_ϵ is not empty if ϵ is small enough).

(ii) *(Useful for bifurcation analysis with λ fixed) Assume that $\tilde{W}''(\lambda,0) < 0$. Then the set*

$$B_\lambda := \{\epsilon > 0 \mid \dim\ker D_u F_\epsilon(\lambda, 0) > 0\}$$

is countable and has the following structure:

$$B_\lambda = \{\epsilon_k \mid k \in \mathbb{N}_0\},$$

where $\epsilon_k = \epsilon_k(\lambda) > 0$ denotes the unique solution of (2.15) corresponding to λ and μ_k. For fixed k, $\epsilon_k(\lambda)$ is strictly decreasing in λ. Furthermore, for fixed λ, ϵ_k is strictly decreasing in k and $\epsilon_k \to 0$ as $k \to \infty$.

Proof. One easily checks that the solution space of (2.14) contains the span of L. In order to prove the other inclusion, observe that the functions v_k form a complete orthonormal system in $L^2_r(B_R(0))$ so that any $v \in X \subset L^2_r(B_R(0))$ can be written as

$$v = \sum_{j=0}^{\infty} \alpha_j v_j \quad \text{in } L^2(B_R(0)),$$

with suitable Fourier coefficients $\alpha_j \in \mathbb{R}$, $j \in \mathbb{N}_0$. By taking the L^2 scalar product of equation (2.14) with the function v_k for a fixed $k \in \mathbb{N}_0$ and integrating by parts to move all derivatives onto v_k, we get

$$\begin{aligned} 0 &= \sum_{j=0}^{\infty} \alpha_j \left\langle v_j\,,\, \epsilon\Delta^2 v_k - \tilde{W}''(\lambda,0)\Delta v_k + G''(0) v_k \right\rangle_{L^2} \\ &= \alpha_k \left[\epsilon\mu_k^2 + \tilde{W}''(\lambda,0)\mu_k + G''(0)\right] \langle v_k\,,\, v_k\rangle_{L^2}\,. \end{aligned}$$

Thus either $\alpha_k = 0$, or μ_k satisfies (2.15).

Concerning the remaining assertions first remember that $\tilde{W}''(\lambda,0) = \lambda^2 + \tilde{W}_0''(0)$ due to (W_{sym}) and $G''(0) \leq 0$ by virtue of (G_1). For fixed ϵ and λ, (2.15) is a quadratic equation in μ_k. Since its constant term $G''(0)$ is less or equal to zero, it has exactly one nonnegative solution so that L contains at most one element. As to (i) and (ii), consider (2.15) solved with respect to λ^2 or ϵ:

$$\lambda^2 = -\frac{1}{\mu_k} G''(0) - \epsilon\mu_k - \tilde{W}_0''(0), \text{ respectively,} \tag{2.16}$$

$$\epsilon = -\frac{1}{\mu_k^2} G''(0) - \frac{1}{\mu_k}\left(\lambda^2 + \tilde{W}_0''(0)\right). \tag{2.17}$$

The assertions of (i) and (ii) are derived from the following observations: The right hand side of (2.16) is strictly decreasing in μ_k and ϵ. Furthermore it eventually becomes negative for large ϵ or μ_k. In particular, since μ_k is increasing in

k and $\mu_k \to \infty$ as $k \to \infty$, there is a number $M = M(\epsilon) \in \mathbb{N}_0$ such that (2.16) is solvable in λ if and only if $0 \le k \le M-1$. Here, $M(\epsilon) \to \infty$ as $\epsilon \to 0$ if $\tilde{W}_0(0) = 0$ and $G''(0) < 0$ or if $\tilde{W}_0(0) < 0$. Likewise, the right hand side of (2.17) is strictly decreasing in λ, and it also strictly increases as μ_k decreases provided that $\lambda^2 + \tilde{W}_0(0) < 0$. □

In particular, we have shown that the dimension of the kernel of $D_u F_\epsilon(\lambda, 0)$ is always less or equal to one. If F_ϵ is of class C^2 then the bifurcation theorem for one–dimensional kernel of Crandall and Rabinowitz [10] is applicable:

Proposition 2.3.6. *(Local bifurcation of a curve of nontrivial solutions) Assume that* (W_λ), (W_0)*–*(W_2), (W_{sym}) *and* (G_0)*–*(G_2) *hold and that* W_0 *and* G *are of class* C^3. *Then we have the following:*

(i) Fix $\epsilon > 0$, *and let* B_ϵ *denote the set of (possible) bifurcation points defined in Proposition 2.3.5, part (i). Then for each* $\lambda_k \in B_\epsilon \setminus \{0\}$ *there exists a neighborhood* $U \subset \mathbb{R} \times X$ *of* $(\lambda_k, 0)$ *and a curve* $s \mapsto \gamma(s) := (\lambda(s), u(s))$, $(-\delta, \delta) \to U$ *of class* C^1 *with the following properties:*

$$\gamma(0) = (\lambda(0), u(0)) = (\lambda_k, 0), \quad u(s) \neq 0 \text{ whenever } s \neq 0,$$
$$F_\epsilon(\lambda(s), u(s)) = 0 \quad \text{for every } s \in (-\delta, \delta), \text{ and}$$
$$\text{the range of } \gamma \text{ contains all nontrivial solutions of } F_\epsilon(\lambda, u) = 0 \text{ in } U.$$

Furthermore, the tangential vector $\dot{u}(0)$ *is equal to the eigenfunction* v_k *which spans the kernel of* $D_u F_{\epsilon_k}(\lambda, 0)$.

(ii) Fix $\lambda > 0$, *and let* B_λ *denote the set of (possible) bifurcation points defined in Proposition 2.3.5, part (ii). Then for each* $\epsilon_k \in B_\lambda$ *there exists a neighborhood* $U \subset \mathbb{R}^+ \times X$ *of* $(\epsilon_k, 0)$ *and a curve* $s \mapsto \gamma(s) := (\epsilon(s), u(s))$, $(-\delta, \delta) \to U$ *of class* C^1 *with the following properties:*

$$\gamma(0) = (\epsilon(0), u(0)) = (\epsilon_k, 0), \quad u(s) \neq 0 \text{ whenever } s \neq 0,$$
$$F_{\epsilon(s)}(\lambda, u(s)) = 0 \quad \text{for every } s \in (-\delta, \delta), \text{ and}$$
$$\text{the range of } \gamma \text{ contains all nontrivial solutions of } F_\epsilon(\lambda, u) = 0 \text{ in } U.$$

Furthermore, the tangential vector $\dot{u}(0)$ *is equal to the eigenfunction* v_k *which spans the kernel of* $D_u F_{\epsilon_k}(\lambda, 0)$.

Remark 2.3.7. If W and G are of class C^4 (and thus $F_\epsilon(\lambda, \cdot)$ is of class C^3), the derivatives of $\lambda(s)$ and $\epsilon(s)$ at $s = 0$ exist up to second order. They can be evaluated explicitly, which gives us a notion of what the bifurcating path must look like, locally. In particular, if we assume that $G'''(0) = 0$ and $D^3 W(\lambda, 0) = 0$

(so that $D_u^3 E_{\epsilon,\lambda}(0) = 0$ for arbitrary λ and ϵ) then the calculations yield that $\dot\lambda(0) = 0$, $\dot\epsilon(0) = 0$,

$$\ddot\lambda(0) = -\frac{1}{3}\frac{D_u^4 E_{\epsilon,\lambda(0)}(0)[\dot u(0)]^4}{D_{\lambda uu} E_{\epsilon,\lambda(0)}(0)[\dot u(0)]^2}, \quad \text{and}$$
$$\ddot\epsilon(0) = -\frac{1}{3}\frac{D_u^4 E_{\epsilon(0),\lambda}(0)[\dot u(0)]^4}{D_{\epsilon uu} E_{\epsilon(0),\lambda}(0)[\dot u(0)]^2}.$$

Due to (W_λ) and (W_{sym}) we have that

$$\begin{aligned} D_u^4 E_{\epsilon,\lambda}(0)[\dot u(0)]^4 &= \tilde W_0^{(4)}(0) \int_{B_R(0)} |\nabla \dot u(0)|^4 \, dx \\ &\quad + G^{(4)}(0) \int_{B_R(0)} |\dot u(0)|^4 \, dx, \\ D_{\lambda uu} E_{\epsilon,\lambda}(0)[\dot u(0)]^2 &= 2\lambda \int_{B_R(0)} |\nabla \dot u(0)|^2 \, dx \text{ and} \\ D_{\epsilon uu} E_{\epsilon,\lambda}(0)[\dot u(0)]^2 &= \int_{B_R(0)} |\Delta \dot u(0)|^2 \, dx. \end{aligned}$$

Thus $\ddot\lambda(0)$ has the opposite sign of $\lambda(0)$ and $\ddot\epsilon(0) < 0$ if $D_u^4 E_{\epsilon,\lambda}(0)[\dot u(0)]^4$ is positive, which is the case for example if the fourth derivative $\tilde W_0^{(4)}(0)$ of $\tilde W_0$ at 0 is positive and $G^{(4)}(0) \geq 0$. Here, note that $\tilde W_0^{(4)}(0) \geq 0$ is necessary for (W_3) since $\tilde W_0^{(3)}(0) = 0$ (remember that $\tilde W_0$ is an even function).

Proof of Proposition 2.3.6. In both cases, we only have to verify the transversality condition of [10]. Since v_k is the k-th radially symmetric eigenfunction of the negative Laplacian with eigenvalue μ_k as in Proposition 2.3.5, we infer from (2.13) that

$$D_{\lambda u} F_\epsilon(\lambda_k, 0) v_k = \frac{2\lambda_k}{\epsilon \mu_k} v_k$$

and

$$D_{\epsilon u} F_{\epsilon_k}(\lambda, 0) v_k = -\frac{1}{\epsilon_k^2 \mu_k^2}\left[\tilde W''(\lambda, 0)\mu_k + G''(0)\right] v_k = \frac{1}{\epsilon_k} v_k$$

by virtue of (2.15). Thus it is enough to show that v_k is not in the range of $D_u F_{\epsilon_k}(\lambda, 0)$ or $D_u F_\epsilon(\lambda_k, 0)$, respectively. This is true because both operators are symmetric with respect to the scalar product in $L^2(B_R(0))$ and v_k spans their kernel. □

If F_ϵ is merely of class C^1, we still have that the Morse index of $-D_u F_\epsilon(\lambda, 0)$ changes by 1 as λ or ϵ, respectively, cross a bifurcation point λ_k or ϵ_k, as we show below. Here, for any continuous linear map $A : X \to X$ whose spectrum does not intersect imaginary axis, the Morse index $m(A)$ denotes the number of eigenvalues of A in $\mathbb{C}^+ := \{z \in \mathbb{C} \mid \operatorname{Re} z > 0\}$, counted by algebraic multiplicity. In our setting, the spectrum of $A := -D_uF_\epsilon(\lambda, 0)$ is real and consists of isolated and geometrically and algebraically simple eigenvalues (except -1 which is the sole accumulation point of the spectrum) and the corresponding eigenfunctions are the radially symmetric eigenfunctions of the Laplacian. Thus $m(-D_uF_\epsilon(\lambda, 0))$ simply is the number of negative eigenvalues of $D_uF_\epsilon(\lambda, 0)$.

Proposition 2.3.8. *Assume (W_λ), (W_0)-(W_2), (W_{sym}) and (G_0)-(G_2).*

(i) Fix $\epsilon > 0$, and let B_ϵ denote the set of (possible) bifurcation points defined in Proposition 2.3.5, part (i). Then $-D_uF_\epsilon(\lambda, 0)$ is invertible for each $\lambda \in \mathbb{R} \setminus B_\epsilon$ and its Morse index is

$$m(-D_uF_\epsilon(\lambda, 0)) = \#\{\gamma \in B_\epsilon \mid \gamma > |\lambda|\},$$

the number of elements of B_ϵ which are larger than $|\lambda|$.

(ii) Fix $\lambda > 0$, and let B_λ denote the set of (possible) bifurcation points defined in Proposition 2.3.5, part (ii). Then $-D_uF_\epsilon(\lambda, 0)$ is invertible for each $\epsilon \in \mathbb{R}^+ \setminus B_\lambda$ and its Morse index is

$$m(-D_uF_\epsilon(\lambda, 0)) = \#\{\gamma \in B_\lambda \mid \gamma > \epsilon\},$$

the number of elements of B_λ which are larger than ϵ.

Proof. (i) The invertibility of $D_uF_\epsilon(\lambda, 0)$ for $\lambda \in \mathbb{R} \setminus B_\epsilon$ is due to Proposition 2.3.5. The number of negative eigenvalues of $D_uF_\epsilon(\lambda, 0)$ can be obtained by counting the numbers $k \in \mathbb{N}_0$ such that $D_uF_\epsilon(\lambda, 0)v_k$ is a negative multiple of v_k, where v_k is the k-th radially symmetric eigenfunction of the negative Laplacian with eigenvalue $\mu_k > 0$ as in Proposition 2.3.5. This amounts to counting the $k \in \mathbb{N}_0$ which satisfy

$$\epsilon\mu_k^2 + [\lambda^2 + \tilde{W}_0''(0)]\mu_k + G''(0) < 0. \tag{2.18}$$

By virtue of the strict monotonicity of the left hand side of (2.18) in λ^2, each such k corresponds to exactly one solution λ_k of (2.15) with $\lambda_k > |\lambda|$, which implies the assertion.

(ii) As before we reduce the assertion to the fact that each $k \in \mathbb{N}_0$ satisfying (2.18) corresponds to exactly one point $\epsilon_k \in B_\lambda$ with $\epsilon_k > \epsilon$, this time using the strict monotonicity of the left hand side of (2.18) in ϵ. □

An immediate consequence of Proposition 2.3.8 is

Corollary 2.3.9. *Assume (W_λ), (W_0)–(W_2), (W_{sym}) and (G_0)–(G_2). Furthermore let $\varepsilon \in \mathbb{R}^+$ and $\Lambda \in \mathbb{R} \setminus \{0\}$ be parameter values such that $D_u F_\varepsilon(\Lambda, 0)$ has nontrivial kernel. Then the index of $F_\epsilon(\lambda, \cdot)$ at the trivial solution changes both if $\epsilon := \varepsilon$ is kept fixed and λ moves across Λ or if $\lambda := \Lambda$ is kept fixed and ϵ moves across ε. More precisely,*

$$|i\,(F_\varepsilon(\Lambda + \delta, \cdot), 0) - i\,(F_\varepsilon(\Lambda - \delta, \cdot), 0)| = 2 \text{ and}$$
$$|i\,(F_{\varepsilon+\delta}(\Lambda, \cdot), 0) - i\,(F_{\varepsilon-\delta}(\Lambda, \cdot), 0)| = 2,$$

whenever $\delta > 0$ is small enough. Here, the index $i(F, x)$ of a compact perturbation of the identity $F : X \to X$ at an isolated zero x of F is defined as the topological degree of F with respect to a suitably small environment of x in X.

Proof. See for example [24]. In particular note that the index of F at an arbitrary zero x (in our case $x = 0$) coincides with -1 to the power of the Morse index $m(-DF(x))$ provided that DF is invertible. □

2.3.2 A global branch of positive solutions

As a consequence of Corollary 2.3.9, the topological index of $F_\epsilon(\lambda, \cdot)$ at the trivial solution changes whenever λ or ϵ (keeping the other parameter fixed) moves across a singular point of the linearization $D_u F_\epsilon(\lambda, 0)$. This of course is the central assumption for the Global Bifurcation Theorem by P.H. Rabinowitz. For the precise statement we first need some notation:

Definition 2.3.10. (i) For fixed $\epsilon > 0$ let S_ϵ denote the closure of the set of all nontrivial solutions of $F_\epsilon(\lambda, u) = 0$, i. e.

$$S_\epsilon := \overline{\{(\lambda, u) \in \mathbb{R} \times X \mid u \neq 0 \text{ and } F_\epsilon(\lambda, u) = 0\}}.$$

For $\Lambda \in B_\epsilon$ (the set defined in part (i) of Proposition 2.3.5),

$$Z_\epsilon = Z_\epsilon(\Lambda) \text{ denotes the connected component of } (\Lambda, 0) \text{ in } S_\epsilon,$$

provided that $(\Lambda, 0) \in S_\epsilon$ (which is the case as we shall see below). In this case, we say that Z_ϵ bifurcates from the trivial solution at $(\Lambda, 0)$. Furthermore, we distinguish two subsets Z_ϵ^+ and Z_ϵ^- of Z_ϵ: If $v \in X$, $\|v\|_X = 1$, spans the kernel of $D_u F_\epsilon(\Lambda, 0)$, then $Z_\epsilon^+ = Z_\epsilon^+(v)$ is defined as the union of the connected components of $Z_\epsilon \setminus (\mathbb{R} \times \{0\})$ which intersect

$$\left\{(\lambda, u) \,\middle|\, |\lambda - \Lambda| < \delta \text{ and } u = sv + \frac{1}{2}sx \text{ with an } s \in (0, \delta) \text{ and } \|x\|_X < 1\right\},$$

for every $\delta > 0$. We call Z_ϵ^+ *the continuum of nontrivial solutions of* $F_\epsilon(\lambda, u) = 0$ *branching off the trivial solution at* $(\Lambda, 0)$ *in direction* v. Likewise, $Z_\epsilon^- = Z_\epsilon^-(v) := Z_\epsilon^+(-v)$ is defined as the continuum of nontrivial solutions of $F_\epsilon(\lambda, u) = 0$ branching off the trivial solution at $(\Lambda, 0)$ in direction $-v$. Note that locally near $(\Lambda, 0)$, any nontrivial solution in Z_ϵ either lies in Z_ϵ^+ or Z_ϵ^-, as a consequence of Lemma 2.3.4.

(ii) For fixed $\lambda \in \mathbb{R}$, the sets S_λ as well as Z_λ (bifurcating at a point $\varepsilon \in B_\lambda$), Z_λ^+ and Z_λ^- are defined analogously, with exchanged roles of λ and ϵ.

Proposition 2.3.11. *Assume* (W_λ), (W_0)-(W_2), (W_{sym}) *and* (G_0)-(G_2).

(i) Fix $\epsilon > 0$. *For each* $\Lambda \in B_\epsilon \setminus \{0\}$, $(\Lambda, 0) \in S_\epsilon$, *and the sets* Z_ϵ^+ *and* Z_ϵ^- *each satisfy one of the following three alternatives:*

(a) Z_ϵ^+ *is unbounded,*

(b) $\overline{Z_\epsilon^\pm}$ *contains a point* $(\tilde{\Lambda}, 0)$, *where* $\tilde{\Lambda} \neq \Lambda$, $\tilde{\Lambda} \in B_\epsilon$, *or*

(c) $Z_\epsilon^\pm$ *contains a pair of points* (λ, u), $(\lambda, -u)$, *where* $\lambda \in \mathbb{R}$ *and* $u \in X \setminus \{0\}$.

(ii) Fix $\lambda > 0$. *For each* $\varepsilon \in B_\lambda$, $(\varepsilon, 0) \in S_\lambda$ *and the sets* Z_λ^+ *and* Z_λ^- *each satisfy one of the following four alternatives:*

(a) $Z_\lambda^\pm$ *is unbounded*

(b) $Z_\lambda^\pm$ *contains a sequence* (ϵ_n, u_n) *such that* $\epsilon_n \to 0$,

(c) $\overline{Z_\lambda^\pm}$ *contains a point* $(\tilde{\varepsilon}, 0)$, *where* $\tilde{\varepsilon} \neq \varepsilon$, $\tilde{\varepsilon} \in B_\lambda$, *or*

(d) $Z_\lambda^\pm$ *contains a pair of points* (ϵ, u), $(\epsilon, -u)$, *where* $\epsilon > 0$ *and* $u \in X \setminus \{0\}$.

Proof. The assertion follows from an refinement of Rabinowitz's theorem on global bifurcation, see [24], II.5.2. Remember that by Corollary 2.3.9, the topological index of the trivial solution changes if the parameter (λ or ϵ) moves across a point where $D_u F_\epsilon(\lambda, 0)$ is singular, and, by Proposition 2.3.5, the kernel of $D_u F_\epsilon(\lambda, 0)$ is one-dimensional at all bifurcation points $\lambda \in B_\epsilon$ or $\epsilon \in B_\lambda$, respectively. □

The remaining part of this subsection is devoted to excluding all but one of the alternatives in each case of Proposition 2.3.11. We are able to do this only for the branch bifurcating at the "first" bifurcation point, i. e. $(\lambda_0, 0)$ in case (i), respectively, $(\epsilon_0, 0)$ in case (ii) (in the notation of Proposition 2.3.5). For (i), the second alternative will be identified as the one remaining, and we will also see that the only other bifurcation point which the branch emanating at $(\lambda_0, 0)$

can reach is in fact $(-\lambda_0, 0)$; in particular both Z_ϵ^+ and Z_ϵ^- contain nontrivial solutions (λ, u) for every $\lambda \in (-\lambda_0, \lambda_0)$. In the case of (ii), we will find that the second alternative has to occur.

By virtue of the a–priori estimates of Corollary 2.2.4 and Proposition 2.2.2, Z_ϵ^+ and Z_ϵ^- are bounded. Also, Z_λ^+ or Z_λ^- are always bounded if they have positive distance to the plane $\epsilon = 0$. The following result shows that the global continuum Z^+ branching off the trivial solution in direction of the first (positive) eigenfunction of the Laplacian cannot meet the trivial solution at another bifurcation point belonging to a different eigenfunction (which, then, changes sign in $B_R(0)$):

Proposition 2.3.12. *Assume* (W_λ), (W_0)–(W_2), (W_{sym}), (W_3) *and* (G_0)–(G_2). *Let Z denote the branch of nontrivial solutions of $F_\epsilon(\lambda, u) = 0$, either bifurcating at $(\lambda_0, 0)$ for fixed $\epsilon > 0$ or bifurcating at $(\epsilon_0, 0)$, for fixed $\lambda > 0$. The corresponding kernel vector v_0 is the first (radial) eigenfunction of the Laplacian; $v_0 > 0$ in $B_R(0)$. Then the part Z^+ of Z branching off in direction of v_0 entirely consists of positive solutions with negative Laplacian, i. e., points $(\lambda, u) \in \mathbb{R} \times X$ respectively $(\epsilon, u) \in \mathbb{R}^+ \times X$ such that $u > 0$ and $\Delta u = u'' + (N-1)r^{-1}u' < 0$ in $B_R(0)$. Furthermore $u' := \partial_r u < 0$ in $\overline{B_R(0)} \setminus \{0\}$, and $\partial_r \Delta u > 0$ on $\partial B_R(0)$.*

The proof relies on Hopf's maximum principle. It follows the lines of a classical argument first used by P. H. Rabinowitz in the setting of scalar semilinear elliptic equations of second order. However the application of the maximum principle is more delicate in our case, because we have to deal with a fourth order equation which is not very well suited for this purpose. We are able to overcome this problem only by using qualitative properties of the solutions arising from the radial symmetry. This poses the fundamental obstruction to an application of our approach to related equations with less symmetry or no symmetry at all.

Proof of Proposition 2.3.12. We restrict ourselves to the case of fixed $\epsilon > 0$; the proof for fixed λ is completely analogous. Define the cone

$$P := \left\{ u \in X \,\middle|\, \begin{array}{l} u > 0 \text{ in } B_R(0),\ \Delta u < 0 \text{ in } B_R(0), \\ \partial_r u < 0 \text{ in } \overline{B_R(0)} \setminus \{0\} \text{ and } \partial_r \Delta u > 0 \text{ on } \partial B_R(0) \end{array} \right\}.$$

We want to prove that $(\mathbb{R} \times P) \cap Z^+ = Z^+$. W.l.o.g., assume that Z^+ is connected; otherwise, the reasoning below can be carried out separately for each component of Z^+. Then it is enough to show that $(\mathbb{R} \times P) \cap Z^+$ is nonempty and both open and closed in the set of nontrivial solutions $S_\epsilon \setminus (\mathbb{R} \times \{0\})$ with respect to the trace topology of $\mathbb{R} \times X$.

First note that the topology of X is strong enough to ensure that P is open in X. Since Z^+ is open in $S_\epsilon \setminus (\mathbb{R} \times \{0\})$ by definition we infer that $(\mathbb{R} \times P) \cap Z^+$ is open in $S_\epsilon \setminus (\mathbb{R} \times \{0\})$.

Second, we observe that $(\mathbb{R} \times P) \cap Z^+ \neq \emptyset$: As a consequence of Lemma 2.3.4, the function u of any point $(\lambda, u) \in Z^+$ which is close enough to the bifurcation point $(\lambda_0, 0)$ in an element of P because $u/\|u\|_X$ is close to the eigenfunction v_0 which lies in the interior of the cone P with respect to the topology of X.

Last but not least, we have to show that $(\mathbb{R} \times P) \cap Z^+$ is closed in $S_\epsilon \setminus (\mathbb{R} \times \{0\})$. For this purpose let (λ_n, u_n), $n \in \mathbb{N}$, be a sequence in $(\mathbb{R} \times P) \cap Z^+$ such that $\lambda_n \to \lambda \in \mathbb{R}$ and $u_n \to u \in S_\epsilon \setminus (\mathbb{R} \times \{0\})$. We claim that $u \in P$ and thus $(\lambda, u) \in (\mathbb{R} \times P) \cap Z^+$, since Z^+ is closed in $S_\epsilon \setminus (\mathbb{R} \times \{0\})$. Obviously, the limit function u is always an element of the closure of P in X, so that

$$u \geq 0, \ \Delta u \leq 0 \text{ and } u' = \partial_r u \leq 0 \text{ in } B_R(0).$$

Since $u \not\equiv 0$, Hopf's maximum principle immediately yields that $u > 0$ in $B_R(0)$ and $\partial_r u < 0$ on $\partial B_R(0)$. Zeros of $u' = \partial_r u$ on $(0, R)$ can be excluded with the help of the Euler–Lagrange equation in the form (2.4):

$$\epsilon \left[(u')'' + \frac{N-1}{r}(u')' - \frac{N-1}{r^2}u' \right] - \frac{\tilde{W}'(\lambda, u')}{u'}u' = -\int_0^r G'(u(s))\frac{s^{N-1}}{r^{N-1}}\, ds \geq 0.$$

Hopf's maximum principle prohibits that a nontrivial solution $u' \leq 0$ of this equation has an interior zero in $(0, R)$. Here, note that $(u')^{-1}\tilde{W}'(\lambda, u')$ remains bounded near zeros of u', since $\tilde{W}$ is of class C^2 and $\tilde{W}'(\lambda, 0) = 0$. As to zeros of Δu, we use $(EL^s_{r,\epsilon})$:

$$\epsilon \Delta^2 u - \tilde{W}''(\lambda, u')u'' - \frac{N-1}{r}\tilde{W}'(\lambda, u') + G'(u) = 0,$$

or, as $\Delta u = u'' + (N-1)r^{-1}u'$,

$$\begin{aligned} &\epsilon \Delta(\Delta u) - \tilde{W}''(\lambda, u')\Delta u = -G'(u) + e(r), \text{ where} \\ &e(r) := \frac{N-1}{r}\left[\tilde{W}'(\lambda, u') - \tilde{W}''(\lambda, u')u'\right]. \end{aligned} \tag{2.19}$$

Due to (G_1) and (W_3) (as well as (W_λ) and (W_{sym})), we have that $-G'(u) \geq 0$ and $e(r) \geq 0$, so that the right hand side of (2.19) is nonnegative. Since $\Delta u \leq 0$ and $\Delta u \not\equiv 0$ we infer as before that $\Delta u < 0$ in $B_R(0)$ and $\partial_r \Delta u > 0$ on $\partial B_R(0)$, concluding the proof. □

Remark 2.3.13. If radial symmetry of u is not assumed, the Euler–Lagrange equation (EL_ϵ) in its strong form can still be rewritten to an equation resembling (2.19):

$$\begin{aligned} &\epsilon \Delta(\Delta u) - \tilde{W}''(\lambda, |\nabla u|)\Delta u = -G'(u) + e(x), \text{ where} \\ &e(x) := \left[\frac{\tilde{W}'(\lambda, |\nabla u|)}{|\nabla u|} - \tilde{W}''(\lambda, |\nabla u|)\right]\left(\Delta u - \frac{\partial^2}{\partial_r^2}u\right). \end{aligned}$$

For the maximum principle argument, it would be enough to have that the "error term" e is nonnegative. Even in the non-symmetric case this is always true if all the super-level sets of u are convex (provided (W_3) is satisfied), since then the second factor of e is less or equal to zero. For a radially symmetric function which is decreasing in the radial direction, this property obviously holds. Establishing it without exploiting the symmetry assumption would allow a generalization to a non-radially symmetric setting.

We summarize:

Theorem 2.3.14. *Assume (W_λ), (W_0)–(W_2), (W_{sym}), (W_3) and (G_0)–(G_2).*

(i) Fix $\epsilon > 0$ and assume that the set B_ϵ introduced in Proposition 2.3.5 (i) contains the point $\lambda_0 > 0$ corresponding to the positive first (radial) eigenfunction of the Laplacian, v_0. Furthermore let Z_ϵ^+ denote the continuum of nontrivial solutions of $F_\epsilon(\lambda, u) = 0$ branching off $(\lambda_0, 0)$ in direction of v_0. Then Z_ϵ^+ is bounded in $\mathbb{R} \times X$, and $\overline{Z_\epsilon^+}$ contains exactly two trivial solutions, namely $(\lambda_0, 0)$ and $(-\lambda_0, 0)$. In particular, there is a nontrivial solution $(\lambda, u) \in Z_\epsilon^+$ of $F_\epsilon(\lambda, u) = 0$ for every $\lambda \in (-\lambda_0, \lambda_0)$.

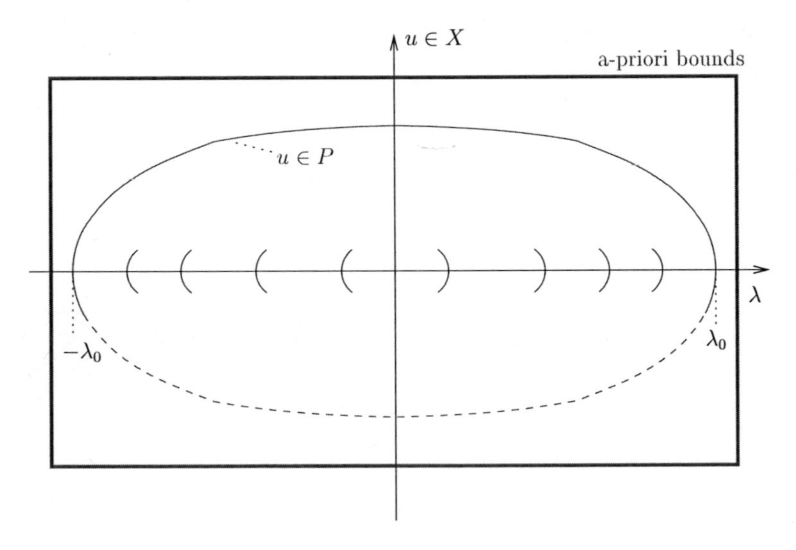

Figure 2.1: Bifurcation diagram for fixed $\epsilon > 0$

(ii) Fix $\lambda > 0$ and assume that $\tilde{W}''(\lambda, 0) < 0$ so that the set B_λ introduced in Proposition 2.3.5 (ii) contains the point $\epsilon_0 > 0$ corresponding to the positive first (radial) eigenfunction of the Laplacian, v_0. Furthermore let Z_λ^+ denote the continuum of nontrivial solutions of $F_\epsilon(\lambda, u) = 0$ branching off $(\epsilon_0, 0)$ in direction of v_0. Then $Z_\epsilon^+ \subset \mathbb{R}^+ \times X$ is bounded, $\overline{Z_\epsilon^+}$ contains no further trivial solution besides $(\epsilon_0, 0)$, and for every $\epsilon \in (0, \epsilon_0)$ there is a nontrivial solution $(\epsilon, u) \in Z_\lambda^+$ of $F_\epsilon(\lambda, u) = 0$.

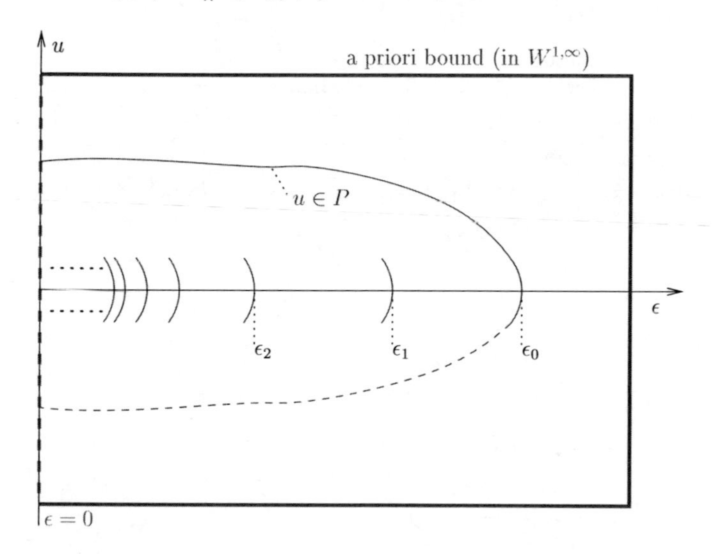

Figure 2.2: Bifurcation diagram for fixed λ where $\tilde{W}''(\lambda, 0) < 0$

Moreover, any function u corresponding to a point $(\lambda, u) \in Z_\epsilon^+$ or $(\epsilon, u) \in Z_\lambda^+$ has the following properties: $u > 0$ in $B_R(0)$, $\Delta u < 0$ in $B_R(0)$, $\partial_r u < 0$ in $\overline{B_R(0)} \setminus \{0\}$ and $\partial_r \Delta u > 0$ on $\partial B_R(0)$.

2.4 The singular limit

As indicated before, with the term "singular limit" we refer to the limit as $\epsilon \to 0$. Our study of this limit is motivated by the fact that a limit u of a sequence of solutions u_ϵ carries the asymptotic qualitative properties of the approximating solutions. In this way we obtain more information on the functions u_ϵ than we are able to derive from (EL_ϵ) directly. Basically, we will prove two things: First, a sequence u_ϵ of solutions of (EL_ϵ) on the positive solutions branch Z_λ^+ is compact in $W_0^{1,P}(B_R(0))$ as $\epsilon \to 0$, for arbitrary $P > 1$. In particular, we get a limit u which satisfies (EL_ϵ) for $\epsilon = 0$, i. e., for every test function $\varphi \in W_0^{1,p}(B_R(0))$,

$$\int_{B_R(0)} \nabla W(\lambda, \nabla u) \cdot \nabla \varphi + G'(u)\varphi \, dx = 0. \tag{EL_0}$$

Since of course u also inherits the sign and monotonicity of the u_ϵ, the results concerning nonnegative critical points of the limit functional obtained in the first chapter can be applied. Second, we show that the limit $u := \lim u_\epsilon$ is nontrivial, i. e. $u \not\equiv 0$ – of course otherwise the previous results would not be useful.

Lemma 2.4.1 (Compactness). *Assume that (u_n), $n \in \mathbb{N}$, is a sequence of radially symmetric functions $u_n \in C^2(B_R(0))$ which is bounded in $W^{1,1}(B_R(0))$. Furthermore assume that $\Delta u_n \leq 0$ for every $n \in \mathbb{N}$. Then the following holds:*

(i) There is a subsequence (u_k) of (u_n) and a radially symmetric function $u \in W^{1,\infty}_{loc}(B_R(0) \setminus \{0\})$ such that both $u_k \to u$ and $u_k' \to u'$ pointwise in $B_R(0) \setminus \{0\}$ and $u_k \to u$ in $W^{1,P}_{loc}(B_R(0) \setminus \{0\}) \cap L^Q(B_R(0))$ for every $P \in [1, \infty)$ and $Q \in [1, N/(N-1))$.

(ii) If the u_n are uniformly bounded in $W^{1,\infty}(B_R(0))$ then $u_k \to u$ in $W^{1,P}(B_R(0))$ for every $P \in [1, \infty)$.

(iii) If the limit u' is continuous on an open set $U \subset B_R(0) \setminus \{0\}$ then $u_k' \to u'$ uniformly on each compact subset $K \subset U$.

Proof. (i) The sign condition on Δu_n implies that

$$\frac{d}{dr}\left(r^{N-1} u_n'(r)\right) = r^{N-1} \Delta u_n(r) \leq 0,$$

in particular, the functions $r \mapsto z_n(r) := r^{N-1} u_n'(r)$ are decreasing on $(0, R)$. As a consequence of this monotonicity, we have that

$$z_n(r) \geq \frac{1}{R-r} \int_r^R z_n(s) \geq -\frac{1}{(R-r)\,\mathrm{Vol}_{N-1}(S^{N-1})} \left\|\partial_r u\right\|_{L^1(B_R(0))}$$

and

$$z_n(r) \leq \frac{1}{r} \int_0^r z_n(s) \leq \frac{1}{r\,\mathrm{Vol}_{N-1}(S^{N-1})} \left\|\partial_r u\right\|_{L^1(B_R(0))} .$$

Thus the bound for u'_n in L^1 entails that z_n is bounded on compact subsets of $(0, R)$. By Helly's theorem (e. g. [28]) we can extract a subsequence (z_k) of (z_n) which converges to a limit z pointwise everywhere on $(0, R)$. Consequently, the sequence u'_k converges pointwise everywhere, too, as well as u_k. Furthermore, the sequence (z_n) is bounded in $W^{1,1}([a, b])$ for every compact interval $[a, b] \subset (0, R)$:

$$\int_a^b |z'_n(r)|\, dr = -\int_a^b z'_n(r)\, dr = |z_n(a) - z_n(b)| \leq C = C(a, b).$$

Thus, by virtue of the compact imbedding, $z_k \to z$ in $L^P([a, b])$ for every $P \in [1, \infty)$, respectively, $u_k \to u$ in $W^{1,P}_{\mathrm{loc}}(B_R(0) \setminus \{0\})$. Moreover, since $W^{1,1}(B_R(0))$ is compactly imbedded into $L^Q(B_R(0))$ for every $Q \in [1, N(N-1)^{-1})$, we also have that $u_k \to u$ in $L^Q(B_R(0))$.

(ii) If the sequence (u_n) is bounded in $W^{1,\infty}(B_R(0))$, we also have that $u'_k \to u'$ and $u_k \to u$ in $L^P(B_R(0))$ for arbitrary $P \in [1, \infty)$ due to Lebesgue's theorem on dominated convergence.

(iii) With an elementary argument based on the monotonicity of the functions z_n one observes that the convergence of (z_k) (and thus also (u'_k) and (u_k)) is uniform on on K using that z (or equivalently, u') is uniformly continuous on K. □

Remark 2.4.2. The limit function obtained in Lemma 2.4.1 is not necessarily of class C^1. Thus one cannot expect convergence in $W^{1,\infty}$. Actually, the (nontrivial) singular limit u of a sequence of u_ϵ on the positive solution branch Z^+_λ never has a continuous first derivative at $r = 0$ whenever $W(\lambda, 0) > \min W(\lambda, \cdot)$, due to Corollary 1.5.4, (1.45). Nevertheless, by the last assertion of Lemma 2.4.1, the convergence of the first derivatives *is* locally uniform on every open set where the limit u' *is known* to be continuous.

A limit of solutions solves (EL_0):

Proposition 2.4.3. *Assume (W_λ), (W_0)–(W_2), (W_{sym}) and (G_0)–(G_2), and let $(\epsilon_n, \lambda_n, u_n) \in \mathbb{R}^+ \times \mathbb{R} \times X$ be a sequence of solutions of $F_{\epsilon_n}(\lambda_n, u_n) = 0$ such that $\epsilon_n \to 0$, $\lambda_n \to \lambda$ and $u_n \to u$ in $W_{1,p}(B_R(0)) \cap L^\infty(B_R(0))$. Then the pair (λ, u) solves (EL_0).*

Proof. By Proposition 2.3.2 $(\epsilon_n, \lambda_n, u_n)$ also solves (EL_ϵ). Integration by parts in the first term yields

$$\int_{B_R(0)} \epsilon_n u_n \Delta^2 \varphi + \nabla W(\lambda_n, \nabla u_n) \cdot \nabla \varphi + G'(u_n) \varphi\, dx = 0,$$

for every test function $\varphi \in C_0^\infty(B_R(0))$. Passing to the limit as $n \to \infty$ yields (EL_0), by a density argument even for arbitrary $\varphi \in W_0^{1,p}(B_R(0))$. □

Proposition 2.4.4 (Nontrivial singular limit). *Assume* (W_λ), (W_0)*–*(W_2), (W_{sym}) *and* (G_0)*–*(G_2)*, and let* $(\epsilon_n, \lambda_n, u_n) \in \mathbb{R}^+ \times \mathbb{R} \times (X \setminus \{0\})$ *be a sequence of solutions of* $F_{\epsilon_n}(\lambda_n, u_n) = 0$ *such that* $\epsilon_n \to 0$, $\lambda_n \to \lambda$ *and* $u_n \to u$ *in* $W_0^{1,p}(B_R(0))$*. Furthermore assume that* $\Delta u_n \leq 0$, $\partial_r u_n \leq 0$ *and* $u_n \geq 0$ *on* $B_R(0)$ *for every* $n \in \mathbb{N}$, $\tilde{W}''(\lambda, 0) \leq 0$ *and* $G''(0) < 0$*. Then* u *does not vanish identically.*

Proof. The proof is indirect. Assume that $u \equiv 0$. In particular, Proposition 2.2.3 and Lemma 2.4.1 imply that

$$u_n \to 0 \text{ in } W^{1,\infty}_{\text{loc}}(B_R(0) \setminus \{0\}) \cap L^\infty(B_R(0)). \tag{2.20}$$

By Proposition 2.3.2 we have that

$$\int_{B_R(0)} \epsilon_n \Delta u_n \Delta \varphi + \tilde{W}'(\lambda_n, u_n')\varphi' + G'(u_n)\varphi \, dx = 0,$$

for every radially symmetric test function $\varphi \in C_0^\infty(B_R(0))$. We integrate by parts in the first term to move the Laplacian onto $\Delta\varphi$ and divide by $\|u_n\|_{1,1} := \|u_n\|_{W^{1,1}(B_R(0))}$ which yields

$$\int_{B_R(0)} \epsilon_n \frac{u_n}{\|u_n\|_{1,1}} \Delta^2 \varphi + \frac{\tilde{W}'(\lambda_n, u_n')}{u_n'} \frac{u_n'}{\|u_n\|_{1,1}} \varphi' + \frac{G'(u_n)}{u_n} \frac{u_n}{\|u_n\|_{1,1}} \varphi \, dx = 0. \tag{2.21}$$

By Lemma 2.4.1,

$$v_n := \frac{u_n}{\|u_n\|_{1,1}} \text{ converges to a limit } v \text{ in } W^{1,1}_{\text{loc}}(B_R(0) \setminus \{0\}) \cap L^1(B_R(0)) \tag{2.22}$$

as $n \to \infty$, at least up to a subsequence. The rest of the proof is organized as follows: We will show that

(a) $v \not\equiv 0$, $v \geq 0$ and $v' \leq 0$ on $(0, R)$,

(b) v satisfies the limit equation

$$\int_{B_R(0)} \tilde{W}''(\lambda, 0)v'\varphi' + G''(0)v\varphi \, dx = 0, \tag{2.23}$$

for every radially symmetric test function $\varphi \in C_0^\infty(B_R(0))$ such that $0 \notin \operatorname{supp} \varphi'$, and

(c) (2.23) does not admit a solution v with the properties (a).

Of course (c) contradicts (a) and (b), concluding the proof.

(a) $v \not\equiv 0$, $v \geq 0$ and $v' \leq 0$ on $(0, R)$:
Clearly, v inherits the sign and monotonicity of u_n. Furthermore we claim that

$$\|u_n'\|_{L^1(B_R(0))} \leq \frac{2N}{R} \|u_n\|_{L^1(B_R(0))}, \text{ for every } n \in \mathbb{N}. \tag{2.24}$$

As a consequence, $\|v_n\|_{L^1(B_R(0))} \geq (2N/R+1)^{-1}$ for every n so that the limit v cannot vanish identically. For the proof of (2.24) define the auxiliary function

$$w_n(r) := -\int_r^R u_n'(s)s^{N-1}\,ds \geq 0, \text{ for } r \in [0, R].$$

Integration by parts yields

$$w_n(r) = u_n(r)r^{N-1} + \int_r^R (N-1)u_n(s)s^{N-2}\,ds,$$

since $u_n(R) = 0$, and by Fubini's theorem, we get

$$\begin{aligned}
\int_0^R w_n(r)\,dr &= \int_0^R u_n(r)r^{N-1}\,dr + \int_0^R \int_r^R (N-1)u_n(s)s^{N-2}\,ds\,dr \\
&= \int_0^R u_n(r)r^{N-1}\,dr + \int_0^R \int_0^s (N-1)u_n(s)s^{N-2}\,dr\,ds \\
&= N\,\mathrm{Vol}_{N-1}(S^{N-1})^{-1} \|u_n\|_{L^1(B_R(0))}.
\end{aligned} \tag{2.25}$$

On the other hand, we obviously have

$$w_n'(r) = u_n'(r)r^{N-1} \leq 0 \quad \text{and} \quad w_n''(r) = r^{N-1}\Delta u_n(r) < 0.$$

In particular, $w_n'(0) = 0$ and thus w_n assumes its maximum at $r = 0$. Since w_n is concave, its integral over $(0, R)$ is at least as big as the area of the triangle with the vertices $(0,0)$, $(0, w_n(0))$ and $(R, 0)$, i. e.,

$$\begin{aligned}
\int_0^R w_n(r)\,dr &\geq \int_0^R \frac{w_n(0)}{R}(R-r)\,dr = \frac{R}{2}w_n(0) \\
&= \frac{R}{2}\mathrm{Vol}_{N-1}(S^{N-1})^{-1} \|u_n'\|_{L^1(B_R(0))},
\end{aligned}$$

which implies (2.24) together with (2.25).

(b) v satisfies (2.23):
For any radially symmetric test function $\varphi \in C_0^\infty(B_R(0))$ such that $\varphi' = 0$ in a vicinity of $0 \in B_R(0)$, we pass to the limit in (2.21) as $n \to \infty$. Using (2.20) and (2.22), this immediately yields (2.23).

(c) (2.23) does not admit a solution v satisfying (a):
Assume that v solves (2.23), $v \geq 0$ and $v' \leq 0$. We claim that this implies that $v \equiv 0$. For any $\delta > 0$ we define φ_δ in such a way that

$$\begin{aligned}&\varphi_\delta \in C^\infty([0,R]),\\&\varphi_\delta = 0 \text{ on } [R-\delta, R],\ \varphi_\delta = 1 \text{ on } [0,\delta], \text{ and}\\&\varphi_\delta > 0 \text{ and } \varphi_\delta' < 0 \text{ on } (\delta, R-\delta).\end{aligned}$$

Now we use φ_δ as a test function in (2.23). Remember that $\tilde{W}''(\lambda, 0) \leq 0$ by assumption and $G''(0) \leq 0$ by (G_1), so that

$$\tilde{W}''(\lambda,0)v'\varphi_\delta' \leq 0 \text{ and } G''(0)v\varphi_\delta \leq 0 \text{ on } (0,R).$$

Thus (2.23) is satisfied only if

$$\tilde{W}''(\lambda,0)v'\varphi_\delta' = 0 \text{ and } G''(0)v\varphi_\delta = 0 \text{ a. e. on } (0,R),$$

for every $\delta > 0$. Consequently,

$$\tilde{W}''(\lambda,0)v' = 0 \text{ and } G''(0)v = 0 \text{ a. e. on } (0,R).$$

Since $G''(0) < 0$, this implies that $v \equiv 0$. □

2.5 The main result

We summarize our results in the main theorem:

Theorem 2.5.1. *Assume (W_λ), (W_0)–(W_2), (W_{sym}), (W_3) and (G_0)–(G_2). Furthermore assume $\tilde{W}_0''(0) \leq 0$ and $G''(0) < 0$ and fix a parameter $\lambda \in \mathbb{R}$ such that $\tilde{W}''(\lambda,0) \leq 0$. Then there exists an $\epsilon_0 > 0$ such that for every $\epsilon < \epsilon_0$, there is a solution $u \in X \setminus \{0\}$ (the set X defined in (2.10), in particular, u is radially symmetric) of (EL_ϵ) such that $u > 0$ in $B_R(0)$, $\Delta u < 0$ in $B_R(0)$ and $\partial_r u < 0$ in $\overline{B_R(0)} \setminus \{0\}$.*

If $(\epsilon_n, \lambda_n, u_n)$ is a sequence with $\epsilon_n \to 0$ and $\lambda_n \to \lambda$ such that, for every $n \in \mathbb{N}$, u_n has the properties listed above, then the sequence (u_n) is relatively compact in $W_0^{1,P}(B_R(0))$ for arbitrary $P \geq 1$. Moreover, any limit u_0 of a convergent subsequence satisfies the following:

(i) u_0 is a nontrivial solution of the limit equation (EL_0).

(ii) u_0 is radially symmetric and $u_0 \in W^{1,\infty}(B_R(0)) \cap W_0^{1,P}(B_R(0))$ for every $P \in [1,\infty)$.

(iii) *$u_0 \geq 0$ in $B_R(0)$, $\tilde{W}'(\partial_r u) \leq 0$ on $B_R(0) \setminus \{0\}$ and $r \mapsto r^{N-1}\partial_r u(r)$ is decreasing on $(0,R)$. Moreover, if $G'(\mu) < 0$ for every $\mu > 0$ then $\partial_r u_0 < -M$ in $B_R(0) \setminus \{0\}$ and $\partial_r u_0(r) \to -M$ as $r \to 0$, where $M := \max\left\{t \geq 0 \mid \tilde{W}(\lambda,t) = \min_{s\geq 0} \tilde{W}(\lambda,s)\right\}$.*

(iv) *If $\tilde{W}(\lambda,\cdot)$ and G satisfy the preliminaries of Theorem 1.6.1, then u_0 is the unique minimum of E_0^λ. In particular, the whole sequence (u_n) converges to u_0, and u_0 does not depend on the choice of the sequence $(\epsilon_n, \lambda_n, u_n)$.*

Proof. The assertions of the first paragraph follow from Theorem 2.3.14 and Proposition 2.3.5. The sequence u_n is bounded in $W^{1,\infty}(B_R(0))$ due to Proposition 2.2.3 and thus relative compact by Lemma 2.4.1. The latter also ensures (ii). Furthermore, (i) was shown in Proposition 2.4.4. Since u_0 solves the limit equation (EL_0), (iii) is a consequence of Corollary 1.5.4. Here, note that $\tilde{W}'(\lambda,t) \geq 0$ for $t \in [-M,0]$ due to (W_3). Finally, Theorem 1.6.1 entails (iv). □

If λ is used as the bifurcation parameter (for fixed ϵ), we even obtain a continuum of solutions in the limit:

Corollary 2.5.2. *Assume (W_λ), (W_0)–(W_2), (W_{sym}), (W_3) and (G_0)–(G_2). Furthermore assume that $\tilde{W}_0''(0) \leq 0$ and $G''(0) < 0$ and let $I \subset \mathbb{R}$ denote the closed set consisting of all $\lambda \in \mathbb{R}$ such that $\tilde{W}''(\lambda,0) \leq 0$ (actually, I is an interval due to (W_3), and $0 \in I$ since $\tilde{W}_0''(0) \leq 0$). Then there exists an $\epsilon_0 > 0$ such that for every $\epsilon < \epsilon_0$, a continuum $Z_\epsilon^+ \subset \mathbb{R} \times (X \setminus \{0\})$ of radially symmetric solutions of (EL_ϵ) branches off the trivial solution at $(\lambda_0, 0)$ and $(-\lambda_0, 0)$ (where $\lambda_0 = \lambda_0(\epsilon) > 0$) in direction of the first positive eigenfunction of the Laplacian. It has the following properties:*

(i) *$\overline{Z_\epsilon^+}$ is connected in the topology of $\mathbb{R} \times X$, $\overline{Z_\epsilon^+} = Z_\epsilon^+ \cup \{(-\lambda_0,0),(\lambda_0,0)\}$, and $I \subset (-\lambda_0, \lambda_0)$.*

(ii) *For each $\lambda \in (-\lambda_0, \lambda_0)$ there is a $u = u_\epsilon \in X \setminus \{0\}$ with $(\lambda, u) \in Z_\epsilon^+$.*

(iii) *If $(\lambda, u) \in Z_\epsilon^+$ then $u > 0$ in $B_R(0)$, $\Delta u < 0$ in $B_R(0)$ and $\partial_r u < 0$ in $B_R(0) \setminus \{0\}$.*

Moreover, the limit set

$$Z_0^+ := \left\{ (\lambda, u) \in R \times W_0^{1,p}(B_R(0)) \;\middle|\; \begin{array}{l} \text{There exists a sequence } (\epsilon_n, \lambda_n, u_n) \text{ with} \\ (\lambda_n, u_n) \in \overline{Z_{\epsilon_n}^+} \text{ for every } n \in \mathbb{N} \text{ such that} \\ \epsilon_n \to 0,\ \lambda_n \to \lambda \text{ and } u_n \to u \text{ in } W^{1,p}, \end{array} \right\}$$

is a connected subset of $\mathbb{R} \times W_0^{1,P}(B_R(0))$ *for every* $P \in [1, \infty)$. *For every* $\lambda \in I$ *there exists a* $u_0 \in W_0^{1,p}(B_R(0)) \setminus \{0\}$ *such that* $(\lambda, u_0) \in Z_0^+$, *and any such function* u_0 *has the properties (i)–(iv) listed in Theorem 2.5.1.*

Proof. Since $W_0''(0) \leq 0$ and $G''(0) < 0$, Proposition 2.3.5 ensures the existence of a first bifurcation point $\lambda_0 = \lambda_0(\epsilon) \in B_\epsilon$ for all sufficiently small ϵ, and Theorem 2.3.14 entails the existence of Z_ϵ^+ and its asserted properties. By virtue of Theorem 2.5.1, the properties of a limit function u_0 claimed in the last sentence of the Corollary hold. Finally, the set Z_0^+ is connected as a consequence of a well known topological argument, see for example [2]. Essentially, its ingredients are the following: One uses (i), (ii) and the fact that each sequence $(\lambda_n, u_n) \subset I \times \tilde{Z}_{\epsilon_n}^+$ with $\epsilon_n \to 0$ has a subsequence converging in $I \times W_0^{1,P}$ for arbitrary P, as already observed in Theorem 2.5.1. Here, Z_0^+ cannot have more than one connected component since the two bifurcation points $(\pm\lambda_0(\epsilon), 0) \in \overline{Z_{\epsilon_n}^+}$ converge as $\epsilon \to 0$. □

Appendix A

A.1 Weierstrass-Erdmann corner conditions

The Weierstrass Erdmann corner conditions are a classical regularity result in the theory of variational problems for functions of one real variable. We give a full exposition here because the literature on this subject is somewhat limited. In particular, the classical result holds for global minimizers only. One slightly more general version – for minimizers relative to a vicinity in the topology of L^∞ – can be found in [19]. It is tempting to state the result for local extrema (local in $W^{1,\infty}$), too, but the proof does not carry over to this situation, cf. [14]. In fact, the problem is due to the use of the topology of $W^{1,\infty}$, since in this space the shift is not a continuous map. Below, we show that one can obtain the corner conditions also for local extrema if an adequate topology is used.

Consider the one-dimensional variational problem given by the functional

$$E(u) := \int_0^1 W(x, u, u')\, dx, \tag{A.1}$$

where $u \in X := W^{1,p}(0,1)$ for a $p \in (1,\infty)$. We could also impose Dirichlet boundary conditions on u without affecting the results below, by restricting X to an appropriate affine subspace. We assume here that

$$\text{(Regularity)} \quad W : [0,1] \times \mathbb{R} \times \mathbb{R} \to \mathbb{R} \text{ is of class } C^1. \tag{A.2}$$

Furthermore, for every $(x, \mu, \xi) \in [0,1] \times \mathbb{R} \times \mathbb{R}$, the derivatives of W with respect to x, u or u' satisfy

$$\text{(Growth)} \quad \begin{aligned} &|W_x(x,\mu,\xi)| \le C(|\mu|)\,(1+|\xi|^p)\,, \text{ and} \\ &|W_u(x,\mu,\xi)| \le C(|\mu|)\,(1+|\xi|^p)\,, \text{ and} \\ &|W_{u'}(x,\mu,\xi)| \le C(|\mu|)\left(1+|\xi|^{p-1}\right), \end{aligned} \tag{A.3}$$

where $C : \mathbb{R}_0^+ \to \mathbb{R}_0^+$ is an increasing function. Due to the assumptions above, E is Fréchet-differentiable in X, and the Fundamental Lemma of Du Bois–Reymond implies that at every critical point $u \in X$, $W'_u(x,u,u')$ is an element of $W^{1,1}(0,1)$ and satisfies the (strong) Euler-Lagrange equation

$$\frac{d}{dx} W_{u'}(x,u,u') = W_u(x,u,u') \text{ for almost every } x \in (0,1).$$

In particular, $W_{u'}(x,u,u')$ is continuous, a property which is called the first Weierstrass-Erdmann corner condition. Classically, the second corner condition is derived by embedding the given functional E into a parametric version of itself which in a sense also allows perturbations of the independent variable x. It is given by

$$\tilde{E}(v,y) := \int_0^1 W\left(y(t), v(t), \frac{v'(t)}{y'(t)}\right) y'(t)\, dt, \tag{A.4}$$

where $(v,y) \in \tilde{X} := X \times \left\{ id + h \mid h \in C_D^1[0,1] \text{ such that } \|h\|_{C^1[0,1]} < 1 \right\}$. It is not difficult to show that $E(u) = \tilde{E}(v,y)$ if $u = v \circ y^{-1}$, where the variables of the two integrals are related by $x = y(t)$.

Theorem A.1.1 (Second Weierstrass-Erdmann corner condition). *Assume (A.2) and (A.3) and suppose that u is a local minimum (maximum) of E in X. Then $(v_u, y_u) := (u, id)$ is a local minimum (maximum) of $\tilde{E}$ in $\tilde{X}$. Furthermore, u satisfies the second corner-condition*

$$W(x,u,u') - W_{u'}(x,u,u')u' \in W^{1,1}(0,1),$$

in particular, the function on the left hand side is continuous. Last but not least, one has the (additional) Euler-Lagrange equation

$$\frac{d}{dx}\left[W(x,u,u') - u'W_{u'}(x,u,u')\right] = W_x(x,u,u'),$$

for almost every $x \in (0,1)$.

The proof is essentially based on the well-known continuity of the shift in L^p ($p < \infty$), which implies that for all functions (v,y) near (v_u, y_u) in $\tilde{X}$, the corresponding non-parametric functions $v \circ y^{-1}$ are near $u = v_u \circ y_u^{-1}$ in X. In particular, local extrema of E always give rise to local extrema of the parametric functional $\tilde{E}$, which is not true in general if the topology of $W^{1,\infty}$ is used instead of $W^{1,p}$.

Lemma A.1.2 (Continuity of the shift in L^p). *For every $w \in L^p(0,1)$, $p \in [1,\infty)$, the map*

$$\eta \mapsto w \circ (id|_{[0,1]} + \eta)^{-1}, \; B_{\frac{1}{2}}(0) \subset C_D^1[0,1] \to L^p(0,1),$$

is continuous at $\eta \equiv 0$.

Proof. We give a short proof for the sake of completeness. First note that whenever $\|\eta\|_{C^1[0,1]} < \delta$ for a $\delta < 1$, the map $id + \eta$ is strictly increasing with first derivative in $(1-\delta, 1+\delta)$. In particular, $id + \eta$ is invertible on $[0,1]$, and the boundary points 0 and 1 of the interval are kept fixed since $\eta(0) = \eta(1) = 0$. Furthermore,

$$(id + \eta)^{-1} \to id \text{ in } C^1[0,1] \text{ as } \eta \to 0 \text{ in } C^1[0,1]. \tag{A.5}$$

To prove continuity of the shift, we approximate w by a sequence of smooth functions $w_n \in C^\infty[0,1]$ such that $w_n \to w$ in $L^p(0,1)$ as $n \to \infty$. Now, let $\epsilon > 0$. We have that

$$\begin{aligned}
&\left\|w - w \circ (id+\eta)^{-1}\right\|_{L^p(0,1)} \\
&\le \|w - w_n\|_{L^p(0,1)} + \left\|w_n \circ (id+\eta)^{-1} - w \circ (id+\eta)^{-1}\right\|_{L^p(0,1)} \\
&\qquad + \left\|w_n - w_n \circ (id+\eta)^{-1}\right\|_{L^p(0,1)} \\
&=: A_1(n) + A_2(n,\eta) + A_3(n,\eta).
\end{aligned}$$

By definition of the w_n, there exists $n_0 \in \mathbb{N}$ such that $A_1(n) < \epsilon$ whenever $n \ge n_0$. Furthermore, by reparametrising the integral we observe that $A_2(n) < 2\epsilon$ provided that $n \ge n_0$ and $\|1 + \eta'\|_{C^0[0,1]} \le 2^p$. Last but not least, $A_3(n_0, \eta) < \epsilon$ if $\|\eta\|_{C^1[0,1]}$ is small enough, due to (A.5) and the uniform continuity of w_{n_0} on $[0,1]$. Summarizing, we have that

$$\left\|w - w \circ (id+\eta)^{-1}\right\|_{L^p(0,1)} < 4\epsilon$$

for sufficiently small $\|\eta\|_{C^1[0,1]}$, which entails the asserted continuity. □

Proof of Theorem A.1.1. We assume that u is a local minimum, the case of a local maximum is completely analogous. First observe that $\tilde{E}$ is well defined in a vicinity of (v_u, y_u) in $\tilde{X}$ and Gâteaux-differentiable at (v_u, y_u) with respect to perturbations in all directions $(\varphi, \eta) \in W^{1,p}(0,1) \times C^1_D[0,1]$. Furthermore, for $h \in \mathbb{R}$ we have that

$$u_h := (v_u + h\varphi) \circ (y_u + h\eta)^{-1} = (u + h\varphi) \circ (id + h\eta)^{-1}$$

is close to u in $W^{1,p}(0,1)$ if $|h|$ is small enough, essentially due to Lemma A.1.2. Since $E(u_h) = \tilde{E}(v_u + h\varphi, y_u + h\eta)$ (as long as h is small enough so that u_h is well defined), this implies that (v_u, y_u) is a local minimizer of $\tilde{E}$. For $\varphi \equiv 0$, differentiation of $\tilde{E}(v_u + h\varphi, y_u + h\eta)$ with respect to h yields the weak Euler-Lagrange equation

$$\int_0^1 W_x\left(y_u, v_u, \frac{v_u'}{y_u'}\right) y_u'\eta + \left[W\left(y_u, v_u, \frac{v_u'}{y_u'}\right) - W_{u'}\left(y_u, v_u, \frac{v_u'}{y_u'}\right)\frac{v_u'}{y_u'}\right]\eta'\,dt = 0,$$

for every $\eta \in C^1_D[0,1]$. Since obviously $u' = v_u'/y_u'$, the Fundamental Lemma of Du Bois–Reymond now immediately implies the assertions. □

Bibliography

[1] Shmuel Agmon. The l^p approach to the Dirichlet problem. *Ann. Scuola norm. sup. Pisa Sci. fis. Mat., III. Ser.*, 16:405–448, 1960.

[2] James C. Alexander. A primer on connectivity. In *Fixed point theory, Proc. Conf., Sherbrooke/Can. 1980*, volume 886 of *Lect. Notes Math.*, pages 455–483, 1981.

[3] Kaushik Bhattacharya. *Microstructure of martensite: why it forms and how it gives rise to the shape-memory effect*, volume 2 of *Oxford series on materials modelling.* Oxford Univ. Press, Oxford, 2003.

[4] P. Celada, G. Cupini, and M. Guidorzi. A sharp attainment result for nonconvex variational problems. *Calc. Var. Partial Differ. Equ.*, 20(3):301–328, 2004.

[5] Pietro Celada. Existence and regularity of minimizers of nonconvex functionals depending on u and ∇u. *J. Math. Anal. Appl.*, 230(1):30–56, 1999.

[6] A. Cellina and G. Colombo. On a classical problem of the calculus of variations without convexity assumptions. *Ann. Inst. Henri Poincaré, Anal. Non Linéaire*, 7(2):97–106, 1990.

[7] A. Cellina and S. Perrotta. On minima of radially symmetric functionals of the gradient. *Nonlinear Anal., Theory Methods Appl.*, 23(2):239–249, 1994.

[8] A. Cellina and S. Perrotta. A correction to the paper "On minima of radially symmetric functionals of the gradient". *Nonlinear Anal., Theory Methods Appl.*, submitted, 2005.

[9] Arrigo Cellina. On minima of a functional of the gradient: Necessary conditions. *Nonlinear Anal., Theory Methods Appl.*, 20(4):337–341, 1993.

[10] M.G. Crandall and P.H. Rabinowitz. Bifurcation from simple eigenvalues. *J. Funct. Anal.*, 8:321–340, 1971.

[11] Graziano Crasta. Existence, uniqueness and qualitative properties of minima to radially symmetric non-coercive non-convex variational problems. *Math. Z.*, 235(3):569–589, 2000.

[12] Bernard Dacorogna. *Direct methods in the calculus of variations*, volume 78 of *Applied Mathematical Sciences*. Springer, Berlin etc., 1989.

[13] Camillo De Lellis. An example in the gradient theory of phase transitions. *ESAIM, Control Optim. Calc. Var.*, 7:285–289, 2002.

[14] Mauro de Oliveira Cesar. Reformulation of the second Weierstrass-Erdmann condition. *Bol. Soc. Bras. Mat.*, 13(1):19–23, 1982.

[15] A. DeSimone, S. Müller, R.V. Kohn, and F. Otto. A compactness result in the gradient theory of phase transitions. *Proc. R. Soc. Edinb., Sect. A, Math.*, 131(4):833–844, 2001.

[16] L.C. Evans and R.F. Gariepy. *Measure theory and fine properties of functions*. Studies in Advanced Mathematics. CRC Press, Boca Raton, 1992.

[17] Gero Friesecke. A necessary and sufficient condition for nonattainment and formation of microstructure almost everywhere in scalar variational problems. *Proc. R. Soc. Edinb., Sect. A*, 124(3):437–471, 1994.

[18] Filippo Gazzola. On radially symmetric minima of nonconvex functionals. *J. Math. Anal. Appl.*, 258(2):490–511, 2001.

[19] M. Giaquinta and S. Hildebrandt. *Calculus of variations 1. The Lagrangian formalism*, volume 310 of *Grundlehren der Mathematischen Wissenschaften*. Springer, Berlin, 1996.

[20] T.J. Healey and H. Kielhöfer. Separation of global solution branches of elliptic systems with symmetry via nodal properties. *Nonlinear Anal., Theory Methods Appl.*, 21(9):665–684, 1993.

[21] W. Jin and R.V. Kohn. Singular perturbation and the energy of folds. *J. Nonlinear Sci.*, 10(3):355–390, 2000.

[22] Hansjörg Kielhöfer. Pattern formation of the stationary Cahn-Hilliard model. *Proc. R. Soc. Edinb., Sect. A*, 127(6):1219–1243, 1997.

[23] Hansjörg Kielhöfer. Critical points of nonconvex and noncoercive functionals. *Calc. Var. Partial Differ. Equ.*, 16(3):243–272, 2003.

[24] Hansjörg Kielhöfer. *Bifurcation theory. An introduction with applications to PDEs*, volume 156 of *Applied Mathematical Sciences*. Springer, New York, 2004.

[25] Hansjörg Kielhöfer. Corrigenda: Critical points of nonconvex and noncoercive functionals. *Calc. Var. Partial Differ. Equ.*, 21:429–436, 2004.

[26] M. Kléman and O. Parodi. Covariant elasticity for smectic–A. *Journal de Physique*, 36:671–681, 1975.

[27] Paolo Marcellini. A relation between existence of minima for non convex integrals and uniqueness for non strictly convex integrals of the calculus of variations. In J.P. Cecconi and T. Zolezzi, editors, *Mathematical theories of optimization, Proc. Conf., Genova 1981*, volume 979 of *Lect. Notes Math.*, pages 216–231, Berlin, 1983. Springer.

[28] I.P. Natanson. *Theorie der Funktionen einer reellen Veränderlichen*, volume VI of *Mathematische Lehrbücher und Monographien. I. Abt.* Akademie-Verlag, Berlin, 1981.

[29] M. Ortiz and G. Gioia. The morphology and folding patterns of buckling-driven thin-film blisters. *J. Mech. Phys. Solids*, 42(3):531–559, 1994.

[30] Jean-Pierre Raymond. Théorème d'existence pour les problèmes variationnels non convexes. *Proc. R. Soc. Edinb., Sect. A*, 107:43–64, 1987.

[31] Jean-Pierre Raymond. Existence of minimizers for vector problems without quasiconvexity condition. *Nonlinear Anal., Theory Methods Appl.*, 18(9):815–828, 1992.

[32] Jean-Pierre Raymond. Existence and uniqueness results for minimization problems with nonconvex functionals. *J. Optimization Theory Appl.*, 82(3):571–592, 1994.

Index